Geometry,
Language and Strategy

K&E Series on Knots and Everything — Vol. 37

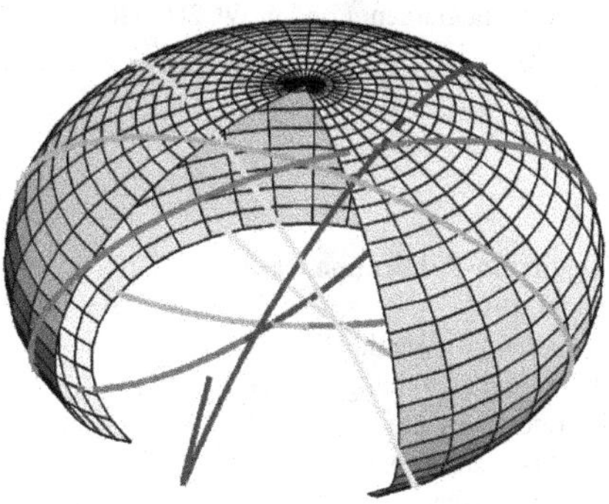

Geometry, Language and Strategy

Gerald H. Thomas

Milwaukee School of Engineering, USA

World Scientific

NEW JERSEY • LONDON • SINGAPORE • BEIJING • SHANGHAI • HONG KONG • TAIPEI • CHENNAI

Published by

World Scientific Publishing Co. Pte. Ltd.
5 Toh Tuck Link, Singapore 596224
USA office: 27 Warren Street, Suite 401-402, Hackensack, NJ 07601
UK office: 57 Shelton Street, Covent Garden, London WC2H 9HE

Library of Congress Cataloging-in-Publication Data
Thomas, G. H. (Gerald Harper), 1942–
Geometry, language, and strategy / Gerald H. Thomas.
p. cm. -- (Series on knots and everything ; v. 37)
Includes bibliographical references and index.
ISBN-13 978-981-256-617-1 (alk. paper)
ISBN-10 981-256-617-1 (alk. paper)
1. Game theory. 2. Statistical decision. 3. Management science. I. Title.
II. Series.

QA269 .T578 2006
519.3--dc22

2005057872

British Library Cataloguing-in-Publication Data
A catalogue record for this book is available from the British Library.

Foreword

This is a most unusual book, breaking new ground in the theory of games.

Thomas begins with the ideas of classical game theory, as formulated by von Neumann and Morgenstern. He takes this approach through a turn that gives rise to a model that has many players, each with their own game parameters and a global structure that partakes of the mathematics of differential geometry. As Thomas says in his Introduction "The best borrowings are probably of a mathematical nature, since Mathematics is superb at speaking deeply without knowing what it is talking about. For purposes of considering dynamics, I believe the ideal mathematical language is geometry." It is the nature of mathematics to have multiple interpretations and applications but it requires artistry and a conceptual shift to find the patterns that match in a new field.

Once there is a multiplicity (imagine indeed a continuum) of players (strategies) in the game, then it is natural to posit a metric that measures the relative distance between two strategies. This book creates, using geometry and physical concepts, a language appropriate to the discussion of the dynamic behaviour of strategies.

The metric in its global properties and its particulars is variable. It is affected by the very strategies that it describes. The space (metric) tells the strategies how to move, while the strategies tell the space how to curve (distort). The situation is exactly analogous to that of general relativity where space tells matter how to move and matter tells space how to curve. Thomas takes this analogy seriously and develops a theory of games that is based in the mathematical and conceptual structure of general relativity and the allied subjects of electromagnetism and gauge theory. The result is a fascinating read that will enlighten game theorists, relativists and all readers with an active curiosity about the structure of the world and the enterprise and exploration that is the making of mathematical models.

Louis H. Kauffman
Professor of Mathematics
University of Illinois, Chicago

Preface

The fundamental job of leaders and managers in industry, I am told, is to determine strategy. Yet leaders and managers at best practice this as an art, not a science. Given the relative importance of strategy to leading and managing, any effort to bring discipline to the subject should be welcomed in the business community. This monograph contributes to the solution to this problem in the framework of the theory of games. I suggest a dynamic extension of the theory based on the wealth of experiences we have from geometry and the physical sciences. My goal is an ambitious one: To create a language appropriate to the discussion of the dynamic behavior of strategies as they occur in the real world. It is not only about rational behavior but also about real world behaviors. The goal of this theory is to enhance our understanding of economic behaviors and make them more susceptible to quantitative analysis.

I have been strongly influenced by work done in Systems Dynamics starting with *The Limits to Growth* by the Club of Rome (1972). Systems Dynamics [*Cf.* Senge (1990) and the foreword by J. W. Forrester in Wolstenholme (1990)] describes behaviors of real world systems with both economic and psychological attributes including production, manufacturing, marketing, delivery, burnout, motivation and training. Using such ideas, I characterize games as consisting of complex positive and negative feedback loops that characterize both the static *Theory of Games and Economic Behavior* by Von Neumann and Morgenstern (1944) and the communication between players which lies outside the normal rules of the game.

The positive and negative feedback loops in Systems Dynamics lead to coupled differential equations which can be solved by computer. This is a situation not unlike physics where initial insight was achieved through phenomenology and experiment. At some point however, further progress is made by ordering the phenomenology and experimental results into a theory. This is often done through analogy to existing theories. Such an ordering creates a new set of distinctions, a language. I note the analogy that Systems Dynamics Models themselves look exactly like interacting springs with various possible connections and spring constants; just another way to think of coupled differential equations. Since such systems can be modeled as "fluids" with elasticity, viscosity, thermal conductivity *etc.*, I propose to write a theory of games as such a fluid.

I have thought about these ideas for a significant period of time, during which I have profited from conversations with many people. I am particularly grateful to Jack Behrend who steadfastly maintained that the concepts in this monograph are important and might be accessible to a CEO and encouraged me to get my ideas on paper in a simple and clear fashion. I owe a debt of gratitude to Lisa Maroski and Helen Kessler who held me to the "impossible promise" of seeing this through completion. More recently, I have enjoyed fruitful discussions with Hector Sabelli and members of his "philosophy group" concerning how the ideas of this monograph might be extended. In preparing the manuscript I have enjoyed the support of the Milwaukee School of Engineering where some of the manuscript has been prepared, my daughter Liisa Thomas for her always valuable legal advice, Steve Wolfram and the Wolfram Research, Inc. and their Partnership Program for support with Mathematica® and from the editor of this series, Lou Kauffman for encouragement. Finally, I am indebted to my wife Kathleen whose patience has made it possible to contemplate and ultimately complete this project and who also has contributed to the improvement of the look of the final manuscript applying her editorial and publishing experience. The superb help and support both mentioned and unmentioned in no way shields me from blame for any faults that remain.

G. H. Thomas

Contents

Chapter 1

Introduction

I have been told that the fundamental job of leaders and managers is to determine strategy. Yet leaders and managers at best practice this as an art, not a science. Those that have gone to business school undoubtedly have learned about the theory of games, not to mention many other theories on how to make strategic decisions. However, I have often heard such managers refer to less scientific or economic books such as *The Art of War* [Sun Tzu (1988)] as the source for their actions rather than any economic theory. Contrast this with a discipline such as Civil Engineering. Not only are the practitioners schooled in a variety of sciences about their craft (Physics, Mathematics and Engineering) but they continue to practice that science throughout their professional careers. Given the relative importance of strategy to leading and managing, I believe any effort to bring discipline to the subject will be welcomed in the business community.

The problems are deep. There is, in fact, no real science; strategy like war *is* an art. There is not even a good language for discussing the subject. The good news is that many people practice this art and so the number of stories in the field is large, reflecting an ever increasing wealth of practical experience. I believe that a rational framework within which to hold such conversations is a contribution to this field. Von Neumann and Morgenstern (1944) provided a significant start at such a framework, though with a static theory. The obvious next step is an inquiry into an extension of their theory to a dynamic theory. There are dynamic theories[1] but none that I know of that make use of the wealth of experience that we have from the physical sciences on how to make such extensions. The problem is how to borrow from those domains. Straight metaphor misses the wealth of experiences with economic behavior[2]. The best borrowings are probably of a mathematical nature, since mathematics is superb at speaking deeply without knowing what it is talking about. For dynamics, I believe the ideal mathematical language is geometry.

My goal is an ambitious one: To create using geometry and the physical sciences, a language appropriate to the discussion of the dynamic behavior of strategies as they occur in the real world. It is not only about rational behavior but also about real world behaviors. The goal is to construct a language, a sufficiently refined language with distinctions that provide insight allowing practitioners to move from an art form to an engineering form[3].

The ***language*** of engineering is mathematics, which consists of marks, strings of characters that brand an idea. Such characters obey rules of their own and can be properly called a language. They form a sharp set of distinctions which I believe will cut through the difficulties of strategic thought. The language of business consists of words. They provide the vernacular, a way to talk publicly about what is observed. I identify such ***distinctions*** in bold–italic in the text that follows and also provide references to them in the index. Both aspects of the language are important, though the business aspect will be more immediately

---

[1] I have found the following interesting and representative: Luce and Raiffa (1957), Williams (1966), Dresher (1981), Eatwell, Milgate and Newman (1987), Ordeshook (1986), Shubik (1991) and Osborne and Rubinstein (1994).

[2] This field has undergone significant change in the last decades and has perhaps moved away from the game theory approach. I am indebted to Dr. Mark Satterthwaite for providing the following set of "course" text-books on modern economics: Myerson (1991) and Mas-Colell, Whinston, and Green (1995). Our goal is to be able to predict economic behavior, which I hope will generate a rebirth of interest in the game theory approach.

[3] The idea of constructing a new language through distinctions might be familiar to those thinking about design questions and artificial intelligence. See for example Winograd and Flores (1986). In this regard, I owe a philosophical debt to these authors, as well as to Rorty (1989) and Rorty (1991).

accessible. It is the engineering aspect however that yields the possibility of measurement, prediction and the application of the scientific method.

Thus this monograph can be read at multiple levels. An initial reading might well focus on the business aspects, reading the mathematical symbols purely as ***mathematical trade-marks*** whose deeper meaning can be deferred to a later reading[4]. A second reading might focus on the mathematics as a language in and of itself and this would demonstrate the quantitative power of the analysis. To facilitate this, in the introductory chapter I provide a sketch or executive summary of the monograph and introduce the mathematics as merely a set of signposts without a great deal of explanation about their structure. I focus on the business aspects. In later chapters, I reverse the process, coming back to each of the signposts and focusing more on their mathematical structures with explanations about their importance and less on the business distinctions. The more detailed mathematical explanations though important, are not as accessible to a large audience; I put these in appendices. In the final chapter I consider the open issues.

1.1 Geometry of Economic Games

I start with the assertion that the ***plays*** of an economic game form a *geo-metry,* literally earth-measurement. Such a ***Strategic Geometry*** is formed from the space of strategic possibilities, decisions or ***strategy–choices***, enumerated in terms of the ***pure strategies*** allowed each player in a multi-person game, coupled with a concept of ***measure*** for each strategy. Each play is an event representing a finished game with all ***moves*** completed and is represented as a point in space; a sequence of plays in a game describes a curve in this strategic geometry space. A sequence of plays of the game suggests a concept of "time" separating successive plays and some concept of relatedness between neighboring plays. I call this dimension ***time–choice***. To describe the relatedness, I introduce the notion of "elasticity" or "connectedness" between "neighboring" games. It is important to note that such elasticity

[4] I thank Ms. Liisa M. Thomas for the legal meaning of trade-marks; I have made no attempt however to strictly adhere to that legal meaning in this monograph.

represents forces that are normally outside of game theory: They include social behavior such as learning and psychological behavior such as burnout. Curves in the ***choice–space*** of strategy–choices and time–choice reflect movement in the geometry and the dynamics of the elasticity or connectedness. A dynamic theory of games is a theory that provides rules for computing the possible curves. An outcome of the theory will be a set of distinctions that characterize the dynamic behavior of games and hence lead to a deterministic view of the flow of decisions. The theory is expected to evolve through the mechanism of the scientific method.

To introduce the notion of strategic geometry, I extend the game–theory concept of strategy as introduced by Von Neumann and Morgenstern (1944). Game–theory enumerates ***pure strategies*** each of which is a complete plan that specifies what a player will do for each ***move*** under every contingency of what the other players could do and under every chance event that might happen. In general, players do not achieve the optimal payoff by choosing pure strategies; they must choose mixed strategies. Game–theory defines mixed strategies as a weighted sum of these pure strategies, subject to the constraint that the weights add up to unity: The interpretation is that the player chooses the pure strategies randomly, with a probability or frequency for each pure strategy set by the weight for that strategy: the odds. Though the odds may change over time, the static theory identifies the limiting strategy or set of ***equilibrium strategies*** which are both stable and fixed. If the game is played over a long period of time, players that adopt the equilibrium strategy will suffer the minimum loss. This framework of game theory is generally believed to be part of any economic theory, though for games with three persons or more, there is no generally agreed identification of the stable and fixed equilibrium behaviors.

I extend the concept of an equilibrium mixed strategy to a time-dependent strategy–choice, which I also call a ***scalar strategy*** by the assertion that the strategy choices or odds change with time and are ***deterministic***, obeying equations that I will specify. The equations will depend on forces that govern the change and will include both game theory forces and non-game theory forces. The possibility will exist for both stable and unstable behaviors, in addition to the static fixed and stable equilibrium behaviors. A scalar strategy can now be thought of as a set of ***strategy–choices*** for each pure strategy: One chooses an amount that is between zero and infinity for each pure strategy. The relative strategy–choices between different pure strategies set the odds. The strategy–choices, normalized by the sum of all strategy–choices, can be interpreted as determining the probability or frequency one chooses for the play. This normalized strategy–choice makes the connection to the static game–theory mixed strategy.

The geometry is described by this set of scalar strategies $\{x^1, \cdots, x^s\}$, along with the time–choice or ***scalar time*** $t = x^0$ that articulates the meaning of ***causality***: "Events that happen later will not impact events that happen now". We shall see that the concept of "***scalar***" is itself an important distinction that describes how different observers describe the same situation. A ***dynamic play*** consists of a scalar strategy with non-negative strategy–choices associated with each pure strategy. A ***dynamic game*** is a time–sequence of dynamic plays and defines a curve in this choice–space. The points along the curve can be labeled by another scalar, the ***strategic distance*** s. A ***strategic geometry*** is thereby defined by the distance between neighboring points.

The idea of distance or measurement is not treated lightly. In geometry, the power comes from measurement. In the normal field of geometry we have thousands of years of experience documenting measurements. In the field of economics and other related disciplines such as social behavior, project management and organizational dynamics, we have relatively few years of experience measuring. The distinctions introduced here are therefore provisional.

The geometry of games is specified by a metric[5]:

$$ds^2 = g_{ab} dx^a dx^b . \qquad (1.1)$$

---

[5] I use a notation similar to Hawking and Ellis (1973). I find their text readable and many of their results helpful in the game theory context here. I recommend their exposition as additional reading. I provide a word of caution however. In this monograph I have maintained a consistent set of conventions that in fact differs from theirs. Hence in some cases results quoted here will differ slightly from their results. Other references may also use different conventions, hence the importance of sticking to a single convention in this monograph.

The metric is a set of rules for measuring the length between neighboring points or plays whose ***mathematical trade-mark*** is the ***line element***. (*Line element* is so named because it is a short segment of the line or curve of interest between dynamically related plays.) The line element reflects the common notion about very short distances being *Euclidean*, *i.e.* distances are laid out on a flat space. Even on flat or Euclidean space, the coordinates need not be at right angles (orthogonal). For non-orthogonal coordinates, a ***metric*** can be specified that is a set of ***measures*** at each point. With these measures specified at every point, it is possible to compute the ***geodesic***, the shortest distance between any two points. I draw from the literature the information needed.

The value of what has been created so far is the notion that successive plays of an economic game are separated by a computable distance determined by the line element. It becomes possible to discuss dynamic behaviors that converge to equilibrium and those that don't. In fact equilibrium ceases to be the main interest, just as it ceases to be in physical models for phenomena such as turbulence and other non-linear behaviors. The metric, the set of measures at each point, will determine the nature of the behavior.

The idea of a ***metric*** determining dynamics is well known in geometry, since dynamics is the statement that things move along the geodesic. What physics adds to the discussion of geometry is the experience that motion or ***dynamics*** determines the metric through the set of ***sources*** given by a new mathematical trade-mark:

$$R_{ab} - \tfrac{1}{2} R g_{ab} = -\kappa T_{ab} \,. \tag{1.2}$$

Most of us are familiar with Newton's idea that matter is the source of gravitational attraction. We may also be familiar with Maxwell's idea that charged matter is the source of electro-magnetic forces. (I propose below that this force is analogous to the economic force.) Einstein (1952) refined both ideas within the context of geometry to state that the energy-momentum of charged and uncharged matter (represented by the mathematical trade-mark T_{ab}) is the source of gravitational attraction (which Einstein represented by the mathematical trade-mark $R_{ab} - \tfrac{1}{2} R g_{ab}$). The refinement was consistent with the theories of mechanics by Newton and Maxwell to the extent that those theories had been verified[6].

Gravity as described by Newton is a force that is felt instantly at any distance from a given source. The theory of Einstein and the theory of Maxwell however forbid any force to travel faster than the speed of light. In particular, the transformation properties of space-time and the sources of matter dictate the propagation properties of the forces. The result of his analysis is that sources of matter can only propagate their forces to other parts of space-time by means of a ***long-range*** messenger field whose trade-mark is $R_{ab} - \frac{1}{2} R g_{ab}$. The analogous statement of Maxwell is that the sources of charged matter can only propagate their forces to other parts of space-time by means of another long-range messenger field whose trade-mark is the electro-magnetic field F_{ab}. The messengers are long-range in the sense that they propagate at the speed of light. Furthermore, the theory determines the ***messenger field*** for matter entirely in terms of the metric trade-mark g_{ab}. Despite the major conceptual difference, Einstein's theory yields the theories of Newton and Maxwell as a good approximation except for astronomically large masses.

The set of measures that specify the distance between neighboring points, the metric g_{ab}, is thus determined by the motion of the sources of matter. From a purely geometric perspective, the messenger field is determined by the trade-mark R_{ab}, which can be identified with the ***curvature*** of space-time. The curvature and the derived trade-mark $R = g^{ab} R_{ab}$ are both functions of the metric. I adopt the ***economic Einstein's equation***, Eq. (1.2) that *the motion of the sources determines the choice-space-time structure and the choice-space-time structure determines the motion of the sources.*

[6] For specific results from physics, the interested reader might consult the excellent volumes of Feynman (1963) for a quite readable description of any of the classical topics of physics, including fluids, geometry, mechanics and electro–magnetism.

1.2 Market Fluid

It might appear that I have introduced an unnecessary complication into the language of the dynamic theory of games. I recall some real world examples:

- A CEO has a goal of adding value to a product or service. To succeed in commerce, his or her company competes with others for a scarce or limited resource for the consumer.
- A project manager manages resources with a goal of delivering their project on time, within cost and of sufficient quality to satisfy their customer.
- An environmental activist hopes to make better use of the world's resources to achieve a safe and sustainable habitat.

In each of these examples, there is a battle between opposing sides for scarce or limited resources. Alternatively, there is a question of which choice to make, including, though not necessarily, the possibility of taking some risk.

The traditional Theory of Games provides a static language for each of these examples with multiple ***plays*** in a game with two or more ***players*** and a payoff for each possible choice of player ***strategy***. I call this language the ***rules-of-the-game***. The rules-of-the-game extracts the essential battle between opposing sides. I believe any theory of social behavior must include these forces and I will discuss these forces in greater detail in Section 1.4. These forces must be added to the metric forces described above to make a complete ***deterministic*** theory.

However, the rules-of-the-game approach supposes that all other real-world effects average out. In this sense the theory is static (not unlike the study of statics in mechanics as applied, for example, to the construction of a bridge). In certain circumstances, these ignored real-world effects might be important and not ignorable. (In the analogy from mechanics, this might correspond to soldiers marching in-step across the bridge.)

How do I include such real world forces and which ones do I include? One approach to computing dynamic behavior in real world behaviors has been with the tools of systems dynamics using sophisticated computer modeling. Systems dynamics models real world behaviors by interactive systems of positive and negative feedback loops. Such models provide information about the small variations away from expected behavior.

There are many examples that might be so modeled.

- A CEO may have a winning attitude, pursue corporate strategies aggressively and communicate clear values to shareholders and employees.
- To improve the world environment, an activist must consider the political forces as well as the management of the world's resources.
- A project manager must be concerned with the effects of burnout and degree of training of the staff, the impact of management overhead and the effect product–defects have on quality and schedule.

In each case, these new attributes describe *how* the games are played and are widely understood to represent forces that drive games towards or away from equilibrium. These forces influence the time attributes of the game, which is extremely important. A CEO has only so much time to achieve success before funding runs out; an environmental activist must achieve success before political forces exhaust critical resources; and a project manager fails in his or her duty if the project significantly misses the deadline.

I assert that these real world attributes are contained in the sources T_{ab} and describe the ***short-range*** connectedness or elasticity of the "medium" or "material" that is one source of the dynamic aspect of games. These short-range forces are distinct from the messenger forces described in the previous Section. I am led to consider the codification of a ***behavioral medium*** for games. I shift from geometry (thinking about how far distinct points are from each other) to medium (how much stuff there is that generates the geometry). This has been extraordinarily helpful in physics. I believe this will be helpful in economics.

What does this mean for game theory? The theory of games as originally formulated describes a static equilibrium choice, one that characterizes rational behavior. If all parties in the game behave rationally, there is a unique set of strategies for each player. In the current view, this behavior is a single rational direction in space. In reality, games are played in many ways and not all choices correspond to this rational direction. There is some connection between these points, some communication that occurs that lies outside the domain of traditional game theory. This communication is the manifestation of the behavior of the players, not the rules-of-the-game. The role of the new behavioral medium is to represent the short-range aspect of this communication. It influences if and how rational behavior comes about.

Computer models of real world behavior that have been constructed are typically complicated and opaque. Some simplification is necessary to characterize or capture significant attributes of real world behaviors. The above thought picture representing as connected points in space multiple games being played simultaneously is a picture of a ***market fluid***. The multiple games are thought of as market "stuff". A small volume $\mathbf{V}$ or cell in choice–space is filled with a given amount of this stuff. The inverse of this volume per unit stuff is the density of stuff per unit volume, ρ, I call the ***market density***. This cell moves in the choice–space. The cell can be defined so that we follow a given amount of this market stuff, being careful to account for whatever stuff enters or leaves the cell. By definition then, the stuff is neither created nor destroyed. The accounting rules that keep track of any increase or decrease in stuff by the flows into or out of the cell, is called the ***conservation of market density***.

Market density endows a market fluid with inertia, an important attribute in describing how easy or difficult it might be to get the fluid to move. Such properties in physics are influenced by gravity and create gravity. If charged, such fluids create and are influenced by electromagnetic fields. However, this fluid is not like any real world fluid for many reasons, not the least of which is that it resides in a higher dimensional space. As such its properties need not be familiar. A starting point will be to generalize the concept of a "perfect fluid" to this multidimensional space, characterizing the fluid by a small number of distinctions that include mass and charge. The language that results will have more generality than this simple model.

The purpose of this monograph is to articulate in some detail both the consequences that follow from this starting point, as well as indicating how this starting point might be extended and in the process indicate that the approach is quite general.

A "perfect fluid" is a behavioral medium with a market density ρ. Such a medium is called ***elastic*** and has additional properties derived from its ***elasticity*** ε that is a given function of the market density: The fluid has a ***pressure*** p which can be thought of as ***power*** or ***control***, an ***energy density*** per unit volume μ and a ***flow*** V_a. Note that each of the business distinctions comes with a mathematical trade-mark, which obey certain rules. For example, these various mathematical trade-marks are related to each other by rules:

$$\mu = \rho(1+\varepsilon)$$
$$p = \rho^2 \frac{d\varepsilon}{d\rho} \quad . \tag{1.3}$$

The rules determine the form of the source for such fluids and are extended to the multi-dimensional choice–space. The sources for this elastic market fluid determine the energy-momentum trade-mark:

$$T_{ab} = (\mu + p)V_a V_b - p g_{ab} \,. \tag{1.4}$$

As asserted, the sources depend on the pressure p, the density μ and the flow V^a. It is possible to determine the elasticity and market density in terms of the energy density and pressure. I reduce the discussion of possible non-rational and hence dynamic behaviors, to a discussion of the flow of play along particular strategic directions and the ease or sluggishness of that flow dependant on the pressure and energy density of the behavioral medium in that neighborhood. The key result is that there will be a dynamic force proportional to the gradient of the fluid flow which is like its physics counterpart of acceleration.

How does this model represent the more usual business picture? The thought picture of multiple simultaneous games being played is represented by points in the choice–space. The concept that the points have an energy density which results from a market density is a sharpening of the concept that certain strategies have a behavioral ***capital***, a weight, ***strategic-mass*** or confidence, which makes them more important than other choices independent of their game–theory value in determining economic equilibrium. Such effects can be expected to be long-range like the game-theory forces. The strategic-mass might be an articulation of a CEO's stubbornness in pursuing a certain direction considered to be a winning approach. It might be the environmental activist's acknowledgment that certain choices are forced due to political realities. It might help support project manager's belief that output is influenced by behavioral factors, as well as unit productivity.

I believe the ***market density*** distinction is a powerful one; I shall use the term ***strategic-mass*** to represent how much market density is in a region of strategy-space. Capital suggests the accumulation of value based on specific means of production and is particularly useful in a traditional economic context. Strategic-mass is a more general concept with applications to more general decision-making processes. Market density will be more useful in the discussions of the rules relating the mathematical trade-marks and in extracting results from the published literature.

The dynamic theory of games has in addition to forces that depend on the motion or ***flow*** of ***strategic-mass*** a force that depends on the behavioral or ***political pressure*** or what might be properly called management ***control*** or focus. I find this force to be short-range and therefore somewhat different than the other forces. However, the concept is a natural extension of the ideas and thought picture proposed. I show that the economic Einstein equations that govern the flow of strategic-mass yield a form that has both the long-range forces of the messenger fields and this short-range force associated with control.

1.3 Thermodynamics of Games

An interesting observation follows from the rules that hold with the above mathematical distinctions and suggests some additional derived business distinctions. The specification of sources T_{ab} consists of a set of numbers at each point in the space of strategies at each instance of time. One particular set of numbers is called the energy $E = T_{00}$ associated with the medium. It is a meaningful and insightful way of talking about the properties of the media and typically is thought of being either ***mechanical*** or ***thermal***. Based on the mathematical rules, the mechanical energy is a property of the motion of the media moving collectively. The thermal energy is a property associated with some internal ***short-range*** structure of the medium whose exact nature is not necessary for the general discussion.

The mathematical rules that generate these distinctions generalize to behavioral media, so these two types of properties can be distinguished in this media as well. I summarize that the energy can be decomposed into thermal and mechanical components:

$$dE = TdS - pd\mathbf{V}. \tag{1.5}$$

The total energy dE in a small region of space is composed of a ***mechanical*** part $-pd\mathbf{V}$ and a ***thermal*** part TdS. The mechanical part is determined by the pressure and density represented here as the reciprocal $\mathbf{V} = \rho^{-1}$. The mechanical part is therefore derived from the distinctions introduced so far.

The thermal part represents the energy that is not accounted for by the purely mechanical collective motion of the fluid. It is described by two new distinctions: ***entropy*** S and ***temperature*** T. They characterize the internal motion. We are familiar with this distinction between collective and internal behavior in markets. A collective move is when all the games being played move towards a new set of strategies. This is the mechanical motion. Internal moves are small variations or fluctuations that occur between different games, though all about a common point. Such small fluctuations are a type of noise. This is the thermal motion. One reflection of such thermal noise is the number of resources necessary to produce something needed for the market; the more noise, the more bodies or resources are required. Thus for initial consideration, temperature can be thought to be the ***resource*** needed to create a unit of something.

Why are two distinctions needed? In common parlance, what we have introduced is the concept of heat (as a manifestation of thermal energy), determined by both entropy and temperature. A small pan filled with boiling water and a large pan filled with boiling water are at the same temperature but have significantly different amounts of heat (thermal energy). This follows from the mathematical rules for these trade-marks. At each point of space the ***behavioral entropy*** $dS = dE/T + pd\mathbf{V}/T$ is determined by the energy and density. As a result of applying the mathematical rules this relationship determines a unique ***behavioral temperature***.

Though I am led to these concepts by analogy to physics, I prove these concepts from the mathematical rules alone [*Cf.* Appendix A]. This is an important point since the distinctions need to stand on their own in the field of games and be supported by observations of phenomena in that field. I believe these new concepts do make sense in the theory of games. I think an indecisive CEO is indicative of thermal motion. An environmental activist unclear on the correct approach may flip-flop on direction. A project manager recognizes that poor communication within an organization causes un-recoverable loss in terms of output and schedule. There are clearly thermal as well as mechanical behaviors; internal as well as collective.

1.4 Rules-of-the-game

I believe that sources are an important aspect of the dynamic theory. I claim that the motion of sources is determined by the choice-space-time structure and the choice-space-time structure determines the motion of the sources. I distinguished those sources from the static aspects of game theory with the caveat that both types of forces are long-range. Where do the static aspects of game theory come from? Are their similarities related in a deep way to a geometric notion?

At the very least it is generally agreed that a key attribute of the rules-of-the-game is the notion that for each individual, there is a payoff or value based on the individual's choice and the choices of all the other players. Equilibrium strategies that differ by an overall scale change are equivalent and so the existence of an equilibrium strategy implies the existence of a line or direction in choice–space. The rules-of-the-game determines a force based on the payoff for some equilibrium strategy. In terms of theories of the physical world, I show that such a force resembles more that of a magnetic field than a scalar force such as Newton's gravity. A magnetic field sets a direction around which charged particles move, rather than attracting them to or repelling them from a specific point. A necessary consequence of this is that the market fluid is ***charged***.

To approach this and ultimately to justify my assertions, I return to a discussion of geometry and observe that it would be elegant to derive the dynamic theory totally from geometry without the need to introduce any additional sources; the hope would be that sources are determined by geometry. This hope is not unfounded and in particular I will use the idea to generate the sources that describe the rules-of-the-game and demonstrate that the resultant forces are consistent with the Von Neumann's ideas of a static theory.

The foundation for this hope is based on an old concept in physics but not one generally known, that familiar notions of space, time and matter might all have their origins within geometry [Weyl (1922)]. In such a theory, gravitational and magnetic forces are unified. Many investigators have used these ideas to create an interesting "Theory of Everything" that unifies both the short-range and long-range forces. It is a theory that has current popularity[7]. In economic theory, the origin of the geometry (the decision space) is actually clearer than the origin of the sources (market fluid). There is no already known underlying theory for sources that we can rely on. Things are therefore quite different from physics. I turn the thinking around and look more intently at the geometry to see if there is a useful and helpful theory of dynamics that can be inferred from that geometry. This inquiry provides the necessary insight for introducing the appropriate magnetic field.

There is still a need for a behavioral medium to represent the short-range forces. These also contribute as sources for the unified geometry. At the current level of maturity of this approach, it is perhaps too big a jump to hope that all sources can be eliminated into some super-unified geometry.

I turn to geometry and use the fact that unification comes about because certain dimensions of the unified space are *hidden*. I assert that

---

[7] An early version of this theory is by Kaluza and Klein, reviewed in Supplemental Note 23 in Pauli (1958). This work is closer in spirit to what we need than the modern theory of strings or the theory of everything. A survey of the modern theory was given on the BBC and is nicely presented in book form by Davies and Brown (1988). A systematic exposition can be found in Green, Schwarz & Witten (1987).

they might be hidden because of symmetries of the unified metric and certain metric components might masquerade as the unknown sources. If the space has symmetries (such as the earth which is idealized as a spherical ball), then distances between points depend on fewer dimensions (the distance between two points on the earth is the one dimensional length of the great circle, specified by the arc length or angle as measured from the center of earth). The simplification does not change the definition of the distance as expressed by Eq. (1.1). The line element depends on a set of numbers, the metric components g_{ab}. When the rules for the line element mathematical trade-mark are applied, the number of such metric components stays the same.

However, it is possible to consider an equivalent geometry in fewer dimensions, the "small" space rather than the original "large" space. The rules for the number of metric components determine there will be fewer metric components in the "small" space than the "large" space. The missing set of components needs to be accounted for: The rules allow one to introduce sources to replace metric components on a one–for–one basis in such a way as to replicate the original geometry. This is an extraordinary result, though one we shall have to accept for now as fact since the mathematical argument is quite involved. For specific symmetries, we need to extract the nature of these additional sources. Though the general result is of some interest, I focus here on the sources that determine the ***rules-of-the-game***.

I enumerate the total number of dimensions of the space. There is one dimension for time–choice. For each player j there will be n_j dimensions corresponding to the number of pure strategies available to that player. To reproduce the static theory, I introduce one additional but natural and hidden dimension for each player that represents the ***value–choice*** for that player. Each player attempts to maximize their perception of value, which does not influence the metric for the game. I translate this requirement to the ***Value–Choice Hypothesis***: **The metric does not depend on the players' value–choices.**

The consequence of this hypothesis is that the theory is equivalent to a theory with one less dimension for each player, one in which for each player j there is a ***market source***, A_a^j. I note that the associated electro-magnetic field Eq. (1.6) is in fact a subset of the unified curvature field. There is a reason why these messenger fields have similar properties.

In addition, there are ***scalar sources*** γ_{jk} which provide ***structural coupling*** of the market to the players[8]. The mathematical trade-mark–rules relate these sources to the underlying geometry[9] and in particular identify these sources with components of the underlying metric. These source and structural coupling components account for all the unified metric components.

The missing ingredient is why I relate the rules-of-the-game forces to the electro-magnetic field. To relate this result to what one would expect from traditional game theory, I summarize that for any game whatsoever, game theory determines a ***payoff matrix*** that specifies payoffs between pairs of players depending on pure strategies chosen by each player. In this context, a pure strategy for each player is a plan for what that player would do, its ***move***, given what is known at the time of that move, for all possible choices of prior moves or chance moves by all the players up to that point. The payoff matrix determines the payments at the end of a ***play*** of the game after all moves have been made. It is far from obvious that every game can be put into this form and even less obvious that most of what we call economic or social behavior indeed has the form of such a game. I refer the interested reader to the extensive literature [Cf. Footnote 1] for details and to Von Neumann and Morgenstern (1944) for the original proofs. These proofs provide a basis for the theory proposed here.

The product, formed from the payoff matrix and the ***strategy vector*** of one player, determines the payoff to the other player. For such a game at equilibrium, the game may or not be fair and the total sum of payoffs may or may not sum to zero. For a general non-zero sum game, each player has a view of the payoffs and therefore of the payoff matrix. A non-zero sum game can be converted to a zero-sum game by adding an

[8] I borrow the term structural coupling from Winograd and Flores (1986).

[9] The notation and results are provided in a later chapter. There I indicate that the larger space with symmetries has metric components $\gamma_{\mu\nu}$ using Greek letters and the smaller space has metric components g_{ab} using Latin letters. The mathematical rules relate the smaller-space metric components to the larger-space metric components.

artificial player that contributes nothing strategic to the game. We can thus deal with zero-sum games.

I will show below that it is always possible to reframe any zero-sum game as a symmetric and hence necessarily fair game. It is done by creating two artificial players whose strategies consist of the total set of the initial player strategies and one new "hedge" player with a single strategy. It is a theorem that a symmetric game always is represented by an anti-symmetric matrix which is the trade-mark for an electro-magnetic field. For such a zero-sum game I specify the symmetric player's ***game matrix*** F^j_{ab} by its components and the mathematical trade-mark rules that the marks $\{a, \ \ b\}$ denote the strategy–choices for the collection of all players. The game matrix will influence the behavior of other games but should not do it instantaneously. General arguments requiring the effects of the game matrix to propagate smoothly to distant parts of space lead to the following trade-mark for the game matrix:

$$F^j_{ab} = \partial_a A^j_b - \partial_b A^j_a . \tag{1.6}$$

Moreover, these general arguments are consistent with identifying A^j_a with the market source defined above from the "large" unified space.

Though the details need to be made clear and indeed will be made clear later, the general consequences should be clear: The game matrix determines the rules-of-the-game. In fact there are ***economic Maxwell equations*** that say that *the game matrix is determined by the motion of the charged fluid and the motion of the charged fluid is determined by the game matrix.*

Of course this will be modified by the economic Einstein equation, Eq. (1.2). I have source components T_{ab} with contributions from the behavioral fluid (determined by the strategic-mass, pressure (or control) and flow of the fluid) and contributions from the game matrix. These contributions are strictly determined by the rules and will be presented in later chapters. Conceptually, we have a complete theory.

1.5 Economic Justification

There are loose ends or questions readers may still have that require more technical detail about the mathematical trade-marks. First, the static theory of games has been alluded to, though no examples presented. Second, I have alluded to the natural relationship between the symmetric game matrix and the electro-magnetic field but without any details. Third, the relationship between the game matrix and the symmetrized payoff matrix has been alluded to but not specified in any detail. I deal with these now.

I start with an example of a ***symmetric*** game that is representative of the many examples available from the literature [See footnote 1]. I consider two armies fighting each other: the Red and the Blue. Both armies have a symmetric set of choices: Fight from the high ground, fight from the low ground or fight as archers. The one that chooses high ground, defeats the one that chooses low ground but will be vulnerable to attack by archers some of the time. The one that picks low ground defeats the archers by hand to hand combat but will be defeated if the enemy chooses high ground. And the one that chooses archers will defeat the enemy on high ground some of the time but will be defeated some of the time by the enemy on low ground.

The following is a possible (common) payoff matrix reflecting these choices:

$$G = \left(\begin{array}{c|ccc|c} \text{Blue/Red} & \text{High} & \text{Low} & \text{Archers} & \\ \hline \text{High} & 0 & 100 & -25 & 2/7 \\ \text{Low} & -100 & 0 & 50 & 1/7 \\ \text{Archers} & 25 & -50 & 0 & 4/7 \\ \hline & 2/7 & 1/7 & 4/7 & \end{array}\right).$$

The game is zero-sum, symmetric and fair, with the payoffs lying between –25 and 25 units of value. No pure strategy is optimal but a mixed set of strategies yields the equilibrium payoff of zero.

The margin shows the optimal mixed strategy vector for each player:

$$\begin{aligned} \text{Blue} &= \{\tfrac{2}{7} \quad \tfrac{1}{7} \quad \tfrac{4}{7}\} \\ \text{Red} &= \{\tfrac{2}{7} \quad \tfrac{1}{7} \quad \tfrac{4}{7}\} \end{aligned}.$$

The mixed strategies for each player are the same. If Blue chooses high ground every time, Red is capable of choosing archers every time and Blue would lose 25 points per game. This is the worst that could happen to Blue. On repeated playing of the battle/game, choosing the three strategies at random using the above strategy vectors for each player to specify the probabilities gives the best outcome for each player independent of what the other player chooses. By changing the payoffs, the payoff matrix can be changed so that one player gets the advantage. For such generalizations, the mixed strategies for each player need not be the same.

Without the choice of archers, the game would express one of the maxims from the Art of War: Always pick the high ground. The presence of the archers–strategy fundamentally changes the nature of the battle, so much so that picking the high ground is not even the most favored strategy. What is favored is the "new" technology of the archer (this could be guerilla warfare or a variety of modern equivalents). The distinction is the concept of ***mixed strategies***, which take full account of the capabilities of the enemy.

This example highlights the distinctions that result from the rules-of-the-game: payoff matrix, strategy vectors, pure strategies, mixed strategies, fair games and zero-sum and non-zero sum games. Games that involve more than two players evoke additional distinctions: coalitions, discrimination, majority, *etc.* These distinctions are consequences of the payoff matrix and the identification of pure strategies. They reflect some but not all the potential forces that may enter in a dynamic game. I assert that a dynamic theory of games is determined by the forces that result from the rules-of-the-game as specified here, the choice-space-time forces and the short range behavioral forces.

The concept from the static theory of games is that every game can be analyzed, pure strategies and a game matrix identified and a rule specified that determines the mixed strategies. Thus each player has some number of pure strategies to choose from, which dictate behavior under every conceivable move of that player or random chance event during the game. Based on the pure strategy choice for each player, there will be a payoff. In general, the player succeeds not by sticking to a single pure strategy but by picking the strategies at random according to determined frequencies. The resultant mixed strategy produces their optimal strategy. I generalize this mixed strategy in the dynamic theory: Successive choices will now follow deterministically as a result of specific forces.

To go one level deeper requires specifying how the equilibrium strategies are obtained from the game matrix G_{ab}. For a mixed strategy X^a for player one and a mixed strategy Y^a for player two, the value of the game will be the trade-mark $X^a G_{ab} Y^b$. Each player optimizes his or her outcome by varying their strategy. The mathematical result of Von Neumann is that if the first player looks for a maximum and the second player then looks for the minimum, the two will be equal to the same numerical value if the order is reversed.

$$v = \max_X \min_Y \left(X^a G_{ab} Y^b \right) = \min_Y \max_X \left(X^a G_{ab} Y^b \right).$$

The expression determines the game value v and is valid for any zero-sum two-person game matrix with any number of pure strategies. Moreover, the result is equivalent to simultaneously maximizing the following two sets of equations with the game value specified above:

$$\begin{aligned} X^a G_{ab} &\geq v \\ -G_{ab} Y^b &\geq -v \end{aligned}. \tag{1.7}$$

The result is not obvious and the proof was given by Von Neumann and Morgenstern (1944). The first equation says that for player two, the worst that can happen is player one maximizes the outcome, so that the best player two can do is pick a pure strategy that minimizes the outcome; player two need not pick a mixed strategy. The second equation is analogous for player one.

Given any game and the numerical values of the game matrix, the above prescription determines the strategies and game value. If both players have two pure strategies, the prescription is not complicated and the interested reader can find solutions in the references provided in Footnote 1. When players have three or more strategies however, the prescription is more complicated. There are algorithms for solving such game theories, one of which is called linear programming which can be applied to any game and is effective in finding solutions even when the number of strategies is large.

Luce and Raiffa (1957) provide an appendix with a good summary of the method. The essential idea is to insure that the game has a positive value by adding a constant to the game matrix so that all terms are positive and then dividing the unknown value in Eq. (1.7) so that new variables appear on the left:

$$\begin{array}{ll} x^a \geq 0 & x^a G_{ab} \geq 1 \\ y^a \geq 0 & -G_{ab} y^b \geq -1 \end{array} .$$

These are constraint equations for a dual set of optimization problems: Subject to the above constraints the first seeks to minimize $\sum x^a$ by varying the vector x^a and the second seeks to maximize $\sum y^a$ by varying the vector y^a. The strategies are determined by normalizing the resultant vectors so that the sum of each is unity.

The solution is simpler for ***symmetric games*** in which each player sees the same game matrix. Since the game is zero-sum, this means that the game matrix is an anti-symmetric game matrix $G_{ab} = -G_{ba}$. For such games the game value is zero, thus Eq. (1.7) simplifies.

For symmetric games there is also a solution using differential equations due to Von Neumann and summarized by Luce and Raiffa (1957) in one of their appendices:

$$\varphi^a = G_{ab} X_b \geq 0$$
$$X = \sum_a X^a \quad \varphi = \sum_a \varphi^a .$$
$$\frac{dX^a}{dt} = \varphi^a - \varphi X^a$$

Equation (1.7) provides the first inequality and the other trade-marks serve to define the differential equation. The equation was constructed to provide a numerical solution for the strategies for any symmetric game and hence for any anti-symmetric game matrix. For large values of the independent variable t, the sum of the strategies X approaches unity and the sum of the payoffs φ approaches zero. Since all payoffs are non-negative, this provides an algorithm for determining the strategies.

My interest in this equation is that it is a prototype for a dynamic equation, with the static solution obtained as a limiting case. The key ingredient of the equation is that equilibrium strategy is a fixed point of the equation: Both sides of the equation vanish. The equation certainly contains the forces due to the rules-of-the-game. Without a full theory however, there is nothing unique about this equation.

I consider other similar equations with the idea of isolating the economic force. I note that the game matrix appears with the term φ^a, which is the product of the game matrix with the strategy vector. To highlight this ingredient, I propose an equally good equation:

$$\frac{df^a}{dt} = f\varphi^a - \varphi f^a$$
$$\varphi^a = G_{ab} f^b / f \quad .$$
$$f = \sum_a f^a$$

In this case the sum of the strategies f is a constant of the motion and so this looks similar to Von Neumann's equation with $X^a = f^a / f$. I transform this equation by introducing a new strategy vector y^a:

$$\frac{dy^a}{dt} = y\varphi^a .$$

The sum of these new strategies y is neither constant nor constrained to be unity. It can be shown however that the new strategies determine the same frequencies as f^a:

$$\frac{y^a}{y} = \frac{f^a}{f} .$$

The virtue of the choice is that the final form shows clearly the form of the interaction between strategies and the game matrix:

$$\frac{dy^a}{dt} = G_{ab} y^b .$$

This equation is linear in the strategy, so that it also holds for the velocity or flow $V^a = dy^a / dt$:

$$\frac{dV^a}{dt} = G_{ab} V^b .$$

I assert that this equation captures the essence of the economic or ***rules-of-the-game*** forces. The equation admits a solution that grows with time along the equilibrium direction, as well as other rotating solutions that are bounded. The frequency y^a / y approaches the equilibrium strategy. Thus the above equation trade-mark has the key characteristic of Von Neumann's equation. The static solution corresponds to the stable point where the velocity is constant on the left and the vanishing on the right results when the velocity is proportional to the equilibrium solution. The above trade-mark also describes the motion of a charged particle in a magnetic field given by the antisymmetric matrix G_{ab}. This demonstrates the assertion made at the outset that removing the restriction that the strategy vector components add to unity exhibits a rules-of-the-game force that is analogous to that generated by a magnetic field.

The result for the rules-of-the-game holds for any symmetric game. To create a satisfactory theory, this result needs to be generalized to games that are not symmetric. Again, Luce and Raiffa (1957) provide an appropriate appendix due to von Neumann that shows that any game can be put into a symmetric form..

I summarize their argument in terms of the payoff matrix[10] and strategy vectors and formulate the static theory in terms of a symmetric game and hence antisymmetric ***game matrix***. I do this for two players engaged in an arbitrary zero-sum game, with m and n pure strategies respectively. The $m \times n$ ***payoff matrix*** for player one is G. The reduced space (*i.e.* the strategy–choice and time–choice space) will have $n \oplus m \oplus 1$ dimensions. In general, the game matrix will have a ***game value*** v and ***defensive strategies*** represented by (generalized) strategy

[10] I consider the payoff matrix for one player, noting that the argument is to be repeated for each player. If all players share the same sense of value-choice, then an approximation is to represent the value-choice dimensions with a single common value-choice dimension.

vectors X and Y for each player that ensure for each respective player a best case payoff as determined in Eq. (1.7) by the product of the payoff matrix and the respective strategy vector.

I reformulate this two-person game with $m \times n$ strategies into an equivalent symmetric game in $n \oplus m \oplus 1$ dimensions with a composite payoff matrix formed from the original payoff matrix:

$$(F_{ab}) = \left(\begin{array}{c|ccc} & n & m & \text{time} = \text{"0"} \\ \hline n & 0 & -G^T & m_0^{-1} G^T X \\ m & G & 0 & -m_0^{-1} GY \\ \text{"0"} & -m_0^{-1} X^T G & m_0^{-1} Y^T G^T & 0 \end{array} \right). \tag{1.8}$$

I note some differences and similarities between this and von Neumann's formulation. In both cases there is an additional player with what is sometimes called a *hedge* strategy. I identify the hedge strategy with *time*. Von Neumann assumes that the hedge payoffs are unity. They can in general be a multiple of unity and in fact I assume that the multiple is itself proportional to the value of the game. The strategies that occur above are equilibrium strategies, so in general $G^T X$ will be a column vector whose every component is equal or exceeds the game value. A similar statement holds for GY.

The key ingredient for dynamics is that the equilibrium strategy be in the null space of the payoff matrix. In other words the equilibrium strategy times the payoff matrix is zero. For the above form of the payoff matrix, the associated null vector must be:

$$(\sigma^b) = (Y \quad X \quad m_0). \tag{1.9}$$

In addition to the game value and equilibrium strategies, there is a parameter m_0 which I am free to specify outside of the rules-of-the-game. It is not a property of the static solution.

The rules that form the composite symmetric game can be articulated by considering component parts. The composite game consists of two symmetric-players, each with all the choices of both players, as well as a choice in time. Symmetric-player one, choosing a strategy from among the m strategies of player one, plays against time and symmetric-player two, choosing a strategy from among the n strategies of player two.

Symmetric-player one will receive the amount determined from the payoff matrix G and a constant value associated with time, with each player having made the indicated choice. Thus symmetric-player one and symmetric-player two act like player one and two respectively. The payoffs to time account for the fact that the original game need not have a zero game value.

A similar argument shows that there is also the possibility that symmetric-player one and symmetric-player two reverse roles and act like player two and player one respectively. In this case, the payoff is opposite in sign. No payoffs occur when symmetric-player one and symmetric-player two make choices from m (or equivalently both from n). The game consists of two embedded versions of the same game in such a way that the composite game is symmetric, having zero expected value.

Because it is an equivalent game, the defensive strategies for the symmetric-players are determined by the strategies of the original game. Because the game is symmetric, the composite defensive strategies for each player are identical. Finally, if there are only two players, the defensive strategies are static equilibrium strategies. The form for the ***equilibrium strategy*** flow σ^b meets these conditions. Equilibrium is indicated by the fact that the payoff when either symmetric-player picks this strategy is zero: $F_{ab}\sigma^b = 0$. The equilibrium strategy depends on the strategies for player one and player two, plus a new distinction, a ***time scale*** m_0, which determines the weight of the time–choice.

1.6 Dynamic Games

I go from a static game determined by fixed mixed equilibrium strategies to a dynamic theory that is ***deterministic*** with time-dependent mixed strategies. I assume that for paths in the space of strategies that are not along the equilibrium strategy flow, there is a defined value for each player payoff matrix F^j_{ab} as well as a defined value for the flow V^b. Even if the payoff matrix were constant over the whole space, non-zero payoffs $F^j_{ab}V^b$ result when the flow is distinct from the equilibrium strategy. I propose that such non-zero payoffs in fact determine the magnitude of the forces generated by the rules-of-the-game [See Eq. (1.10) below].

The argument is extended to any number of players. With two players, the ***equilibrium strategy*** is also a ***defensive strategy***. For N players, there will be a defensive strategy even if there is no clear notion of what constitutes a stable set of equilibrium strategies. For such games, each player is defensive if he plays the game he sees (using his payoff matrix F^j_{ab}) as if the other players join together in a ***coalition*** against him. With three or more players, many such coalitions are possible: There will be effective two-person games according to each partitioning of the players into disjoint coalitions. An analysis of such coalitions was used by Von Neumann and Morgenstern to determine one definition of stable behavior.

The problem to resolve for games with more than two players will be the form of the game matrix. Although the payoff matrix G between any two players might be constant over space, the choice of strategy vectors that appear in the game matrix depends on the coalition formed, if any. Each coalition specifies a different path in space, which suggests that the game matrix has a different value along each of these paths. The full game matrix is then an interpolation between these known values. There is support for the idea introduced by Von Neumann and Morgenstern (1944) that these more complicated games have a stable pattern of behavior as opposed to a single equilibrium strategy. This pattern of behavior is determined by the behavior along the various coalitions. What I see in addition is that stable equilibrium, if it exists, need not be the set of defensive positions, nor correspond to the stable behavior patterns of von Neumann and Morgenstern.

With these arguments and caveats, I am led to a dynamic theory given by the ***economic Einstein's equation***, Eq. (1.2) that *the motion of the sources determines the choice-space-time structure and the choice-space-time structure determines the motion of the sources.* In particular, the theory determines the motion of the sources and hence the rate at which the flow changes in terms of previously discussed trade-marks:

$$\frac{DV^a}{\partial s} = g^{ab} V_j F^j_{bc} V^c + \left(g^{ab} - V^a V^b\right) \frac{\partial_b p}{\mu + p} - \tfrac{1}{2} g^{ab} V_j V_k \partial_b \gamma^{jk} \,. \quad (1.10)$$

In this monograph, I demonstrate that this form comes from a more technical exposition and articulation of the trade-mark rules of Eq. (1.2). I see the possibility that the rules lead to a quantitative relationship between the flow of the behavioral strategic-mass in terms of the metric field, the rules-of-the-game, fluid dynamic forces based on pressure gradients and structural coupling sources.

The static theory of games is a special case in which the flow of strategic-mass is constant. The dynamic theory of games corresponds to the case that the flow is not constant and its acceleration (the rate of change $DV^a/\partial s$ of the flow) is determined by forces corresponding to rules-of-the-game notions, as well as notions of behavior and possibly additional hidden sources. Though this has the sound of a metaphor from physics, it is not; rather it is a strict result of the rules applied to the relationship between the sources and the metric in choice–space.

I sketch the argument for the special case that recovers the static theory where the rules-of-the-game provide no external forces that act upon the strategic-mass. The static theory of games ignores the influence of hidden sources. All static effects are assumed to be captured by the rules-of-the-game. Still, behavioral sources are reasonably expected to move the flow of strategic-mass if some strategy choices are subject to more pressure (or control) than others: Thus the static theory of games assumes no such pressure differences. Similarly the static theory assumes no differences associated with changes in the hidden sources.

It is then plausible that the rules for the static theory relate the movement of strategic-mass only to the game matrix:

$$\frac{DV^a}{\partial s} = g^{ab} V_j F^j_{bc} V^c .$$

This is the reformulation of the Von Neumann model in Section 1.5. This is the equation for a charged fluid moving in an electro-magnetic field in choice–space, justifying the comment made earlier that the motion looks like that of a particle in a magnetic field. For the special case that the flow takes on the equilibrium value, there are no forces (right-hand side) and there is no acceleration (left-hand side).

I obtain a ***deterministic*** generalization of the static theory. There is acceleration due to non-zero payoffs as seen by each player, generated by the rules-of-the-game. The strength of each player's contribution is determined by the "charge" or ***value–scale*** V_j. Along the equilibrium strategy flow σ^b, the payoff $F^j_{ab}\sigma^b = 0$ vanishes. For a zero-sum game the equilibrium flows for each player are equal. If the equilibrium payoff flow is also constant, then the rule above is satisfied. Of course I need to say much more about the rules that govern the trade-marks to elevate this argument to something other than a sketch. I will illuminate properties of Eq. (1.10) in later chapters.

The acceleration of strategic-mass provides a generalization of the equilibrium strategy: A dynamic strategy is associated with any flow satisfying Eq. (1.10). Moreover, the acceleration of strategic-mass provides a means of calculating the behavior near equilibrium if one exists or more generally near any ***fixed point*** defined as one having zero acceleration and whose sum of forces vanishes. It satisfies the equation for all time and so obeys the rule that the flow is constant and not acted upon by an external force. For the behavior to be ***stable***, I require in addition that for a sufficiently small change away from this point, the resultant flow stays "close by" for all time. If this is not possible, then the behavior is ***unstable***. For example, in a game with three or more persons, the behavior near defensive or coalition points might be unstable. Even for this case however there could be regions with an analogous stable property: Anything that starts near such a region remains nearby. This would provide an example of the ***stable pattern of behavior*** of Von Neumann and Morgenstern (1944).

The acceleration of strategic-mass provides deterministic non-rational behaviors or curves that are reasonable extensions of rational behavior. I take as "reasonable" those extensions that approach rational behavior if the set of strategies exhibit stable behavior. The form above based on the rules-of-the-game forces was introduced by Von Neumann as a "trick" or practical proposal for calculating equilibrium strategies (See Section 1.5). My inquiry was initiated to see if this "trick" could be developed into an interesting theory of dynamics with the fixed points describing the static equilibrium. Such a theory would carry out the next phase of Von Neumann's program. In fact I do obtain an interesting theory, one in which strategic-mass is forced to move not only by rules-of-the-game forces but by behavioral and metric forces.

1.7 Nature of Time

I conclude this introductory chapter with a few comments about the nature of time in a theory of games. I would like time to correspond to the usual notion and imply some type of ***causality*** of events. Things that happen later do not impact things that happen earlier. What is usual? To say that time corresponds to the usual notion ignores the fact that controversies have historically surrounded time. The most recent accepted notions of time come from physics. In this arena, there are two conflicting categories corresponding to alternative geometries, those in which the resultant geometry is named Euclidean and those in which the resultant geometry is non-Euclidean, named Riemannian. I see two fundamental categories occurring for the dynamic theory of games as well. My analysis suggests a preference for what is called Riemannian and rests on the analogy of the magnetic field in physics with the game matrix generating the rules-of-the-game. The magnetic field generates stable behavior for charged particles moving along the field when they are sufficiently nearby. It is precisely this behavior I expect for the flow of strategic-mass near a stable defensive path for a two-person game.

This choice determines the form of the metric. The pure strategies form a space that is locally Euclidean. This means that locally I could choose an orthogonal coordinate system in which the scales are the same for the space (non-time) components. I label the scale with -1. The question raised above about Euclidean or Riemannian geometries is whether the scale for the time component has the same label or one that is the same with an opposite sign. That choice determines what is technically called the ***signature*** of the geometry and is an attribute that is true at every point in space and is the same for every possible observer. It is fundamental to the theory and not an attribute of any specific model calculation. A Euclidean signature is one in which the time component has the same sign as the space components. I choose the opposite sign, namely $+1$ corresponding to a Riemannian signature[11]. With this choice,

---

[11] An equivalent choice is that the space components are $+1$ and the time component is -1. The Riemannian distinction is when the choices between the space and time components are opposite. My choice is an arbitrary convention that leads to no observable consequence. Indeed, the literature in physics is split on this convention.

the game matrix is analogous to the electro–magnetic field in physics. The choice has to be viewed however as provisional, subject to change as more experience is gained with the theory.

1.8 Outline

I have provided a business sketch of a dynamic theory of games focusing more on the business distinctions than the mathematical trade-mark distinctions. As I have remarked, the rules governing the mathematical trade-marks require elaboration to further illuminate the theory and make possible quantitative predictions as the theory is ***deterministic***. I do this in subsequent chapters. There I provide an exposition of the theory and the distinctions related to economic behavior. In other words I fill in the sketch. With economic justifications, I borrow from geometry a richly constructed language and rules and borrow from physics insight and short-cuts in understanding these rules. Such short-cuts provide additional distinctions that can be used in the economic theory.

I suggest that this monograph be read at a number of levels. One level is designed to introduce a new theory that provides an understanding of economics: a dynamic theory of games. This introduction provides that level. On a second level I wish to impart to interested readers sufficient tools for them to compute behaviors from the theory. In the chapters that follow in the main body of the monograph, I have taken an approach long adopted by mathematicians: Try to make the ideas plausible to interested and trusting readers yet provide a guide for replicating and verifying the ideas for those less trusting souls like me who need to verify that the derivations are correct before accepting them as insightful. In the body of the monograph, I acknowledge that for a small number of readers, I may not have provided enough of a guide as to the correctness of the mathematical arguments, even to those with the necessary mathematical sophistication. For those readers I have provided a third level in the appendices.

The goal in this monograph is to advance our understanding of the field of economic interactions by providing a new language based on a new set of distinctions inspired in part by a science metaphor and in part by observations in the field. The power of creating language however is to come to an understanding of "what we don't know we don't know"[12]. I believe all three levels contribute to that goal.

I have organized the monograph generally along the lines of the introduction. In Chapter 2, I define the geometry of games. I introduce some common geometric notions and show how they are expressed in the language of differential geometry. In Chapter 3, I obtain a first form for the acceleration of strategic-mass. An important question of whether thermodynamics make sense in such a theory is answered in the affirmative in Appendix A. In Chapter 4, I inquire into the symmetries of a game and identify key symmetries. I obtain a more usable form for the acceleration of strategic-mass. The discussion relies on notions from differential geometry which are presented in Appendix B. A general and useful class of symmetries, called central symmetries, is defined and expanded upon in Appendix C. I provide a soluble model of the full economic Einstein equations using these ideas in Appendix D, with a numerical example for a fair game in Appendix E. In Chapter 5 I analyze the properties expected for general solutions. I describe the basic behavior when only the rules-of-the-game apply and provide an analysis of the general case with a graphical presentation in Chapter 6. A useful insight into the solutions and their existence is provided by streamlines, presented in Appendix F. I describe a generalization of the perfect fluid in Appendix G. In Chapter 7 I return to the main ideas raised in the introduction about language and suggest applications and open problems that might be treated by the theory.

---

[12] "But if we could ever become reconciled to the idea that most of reality is indifferent to our descriptions of it, and that the human self is created by the use of a vocabulary rather than being adequately or inadequately expressed in a vocabulary, then we should at last have assimilated what was true in the Romantic idea that truth is made rather than found. What is true about this claim is just that "Languages" are made rather than being found, and truth is a property of linguistic entities, of sentences."—Rorty (1989).

Chapter 2

Rules-of-the-Game

The static Theory of Games consists of an analysis of all games consisting of multiple ***players***, multiple ***moves*** (as in chess) and elements of randomness (as in poker). For my purposes, the usefulness of and the main thing I take from the analysis by Von Neumann and Morgenstern (1944), is that all games can be reduced to a statement about the ***number of players***, whether or not the game is ***zero-sum*** and for each player, a "***play***". Each play is chosen from a list of ***pure strategies***, where each pure strategy is an entire specification of what to do by one player for every possible move of all other players for the full duration of the game and for every random contingency that the game depends on. The way a play is chosen can be a pick of a pure strategy or a random pick of any possible pure strategy based on assigned probabilities. The latter is called a ***mixed strategy***. These ideas appear to have been adopted by economists.

I extend the ***equilibrium mixed strategy*** to a set of time-dependent ***strategy–choices*** for each pure strategy, also removing the restriction that the sum of a player's strategy–choices is constant. The set of strategy–choices for all players defines a point in the composite ***choice–space***. Once the choices are made, the outcome is determined and each player receives or pays an amount called the payoff. The ***payoff*** assumes that there is a currency representing an agreed ***value*** between players.

I want to express these ideas using the mathematical trade-marks and in so doing indicate some of the rules governing such trade-marks[13].

---

[13]There are many textbooks on geometry and on the rules alluded to. I quote a couple for the avid student: A more traditional view of geometry is provided by Synge and Schild (1949). A more modern treatment of geometry is Göckeler and Schüker (1987), though

There will be some repetition of the ideas articulated in the introduction and this is purposeful. In part I am introducing the language of the mathematical trade-marks through "speaking", not unlike an immersion class into a foreign language introduces the foreign language through conversation. More importantly, I need to provide justification or at least plausibility that the use of the mathematics is consistent with the economic ideas. By its nature, the mathematics is a condensed language which presupposes a great deal about the terms expressed.

The purpose of this chapter therefore will be to reformulate the ideas of the Introduction with mathematical precision and justify the use of that mathematics. I start by considering how economic concepts transform and show that such transformations dictate the form of the trade-marks, coming to the conclusion that games are covariant. The form of the game matrix is known from the static theory at the defensive position and needs to be extended to the full choice–space. This provides some general attributes of the game matrix. The geometry of the space is characterized by the "shortest paths" that connect points, called geodesics. These paths are identified for choice–space. Because the geometry is not expected to be "flat", the distinction that the geometry is locally flat is introduced. This chapter concludes with the final form of the dynamic game theory hypothesis that reflects the rules-of-the-game.

2.1 Games are Covariant

For a two-person zero-sum game, the outcome of the game is determined by a payoff matrix $\mathbf{G}$, the rate $\mathbf{X}$ player one chooses for each of his pure strategies and the rate $\mathbf{Y}$ player two chooses for each of his pure strategies: $\mathbf{X}^{\mathrm{T}} \cdot \mathbf{G} \cdot \mathbf{Y}$ represents the payout[14] to player one based

---

the latter is not for the light-hearted. In some ways however the modern treatment is, after an initial struggle, conceptually easier and agrees more with Hawking and Ellis (1973). The caution for all of these books is that they differ in their conventions and so identical results don't necessarily appear identical when compared.

[14] Because neither the choices nor their rates are normalized to unity, strictly speaking the result is proportional to the payout. I note that under equilibrium conditions, the strategy–choices grow linearly in time, so the strategy–choices and their rates become effectively proportional after sufficient time.

on the strategy–choices. The mathematical trade-mark $\mathbf{X}^{\mathrm{T}} \cdot \mathbf{G} \cdot \mathbf{Y}$ is a compact notation that provides the numerical value of this outcome:

$$\mathbf{X}^{\mathrm{T}} \cdot \mathbf{G} \cdot \mathbf{Y} = \sum_{i=1}^{m} \sum_{j=1}^{n} X_i G_{ij} Y_j \ .$$

The matrix notation implies a specific summation over the rates $\mathbf{X}$ and $\mathbf{Y}$ of each player. The compact notation for the rates of each player defines a ***vector***, a set of numbers giving the relative rate or weight for each pure strategy. It also defines a game matrix $\mathbf{G}$ that determines the push–pull effect of each player's choices that will result in a payoff to one of the players.

The payoff $\mathbf{X}^{\mathrm{T}} \cdot \mathbf{G} \cdot \mathbf{Y}$ would appear to depend on the way strategies have been defined by those playing the game. This is not so however. A different set of players playing the same game might arbitrarily choose linear combinations of the pure strategies as what they call pure. The payoff does not change as a result of different labels. The distinction is that the payoff is a ***scalar*** quantity; it is the same independent of the linear transformation of the strategy–choices.

Saying this in a more mathematical language, the same game can be viewed by changing the ***basis*** to another system by means of a linear transformation of the coordinates: $\mathbf{X} = \mathbf{S} \cdot \overline{\mathbf{X}}$ and $\mathbf{Y} = \mathbf{T} \cdot \overline{\mathbf{Y}}$. Quantities that obey rules such as these for the player strategies are called ***vector*** quantities and the components are called vector components. The payout will be unchanged as long as the game matrix is transformed in an inverse manner to each of the vectors: $\mathbf{G} = \left(\mathbf{S}^{-1}\right)^{\mathrm{T}} \cdot \overline{\mathbf{G}} \cdot \mathbf{T}^{-1}$. Not only does the value of the payout not change, the form of the payout in terms of the transformed (rates of) strategy–choices and payouts is unchanged:

$$\mathbf{X}^{\mathrm{T}} \cdot \mathbf{G} \cdot \mathbf{Y} = \overline{\mathbf{X}}^{\mathrm{T}} \cdot \mathbf{S}^{\mathrm{T}} \cdot \left(\mathbf{S}^{-1}\right)^{\mathrm{T}} \cdot \overline{\mathbf{G}} \cdot \mathbf{T}^{-1} \cdot \mathbf{T} \cdot \overline{\mathbf{Y}} = \overline{\mathbf{X}}^{\mathrm{T}} \cdot \overline{\mathbf{G}} \cdot \overline{\mathbf{Y}} .$$

The notation therefore expresses a reality about games, that the description has an apparent sameness or ***covariance*** of form under linear transformations. Differences in taste or style might favor one choice of strategies as a basis over another. Such choices however do not lead to a different form of the theory, nor to different results for such important questions as "what are the equilibrium strategies?"

2.2 General Attributes of the Game Matrix

I generalize the discussion to that for N players[15]. I define the ***defensive player flow*** as one in which there is a set of payoff numbers f_{mn} that define the payoff matrix. Between any two players α and β, the payoff matrix is $f_{m(\alpha)n(\beta)}$. In other words, the total set of strategy–choices (a point in the choice–space) of all players represented by the ***indices*** m and n is ***partitioned into disjoint sets*** labeled by the indices α and β that represent the specific player. Based on these numbers, there is a set of flow components $\zeta^{m(\alpha)}$ that represent the rate player α chooses each pure strategy when playing against all other players, including the possibility that the other players might collude and form ***coalitions***: This player is assured of receiving v^{α} based on this scenario. This is the defensive position for this player. In normal play, there is no internal payoff and so elements such as $f_{m(\alpha)n(\alpha)}$ are zero. However there could be internal ***factions*** so I allow at least provisionally the possibility that such payoffs are not always zero. The static theory assumes that the payoff matrix f_{mn} is the same set of numbers when computing each player's defensive position.

With N players there will be a number of partitions of these players into two distinct coalitions and again I use the same set of payoff numbers f_{mn} to compute the payoff to each coalition. To each partition $m(S)$ of the full set of strategies m there will correspond an effective two-person game, one played against the complementary partition $m(-S)$. Each side of the partition chooses at the rates $\bar{\zeta}^{m(S)}$ and $\bar{\zeta}^{m(-S)}$ respectively. The set of all possible coalition payoffs $\bar{\zeta}^{m(S)} f_{m(S)n(-S)} \bar{\zeta}^{n(-S)}$ is determined from the original payoff matrix f_{mn}. In this way an effective two-person game payoff matrix $f_{m(S)n(-S)}$ is formed. For each partition there will be ***defensive coalition flows***, $\zeta^{m(S)}$ and $\zeta^{m(-S)}$ that represent optimal strategic rates for the coalitions in the effective two-person game. Thus in addition to the defensive player flows, there will be defensive coalition flows which might dictate dynamic behavior.

---

[15] In this discussion, I consider the games from the point of view of player α_0. To obtain somewhat simpler expressions, I don't indicate the dependence on this player. To obtain the full expressions, replace the payoff matrix f_{mn} with $f_{mn}^{\alpha_0}$.

To gain insight into the possibilities involved with multiple players, it is useful to consider the payoffs f_{mn} as having the same value at every point in the strategic space and at all points in time. This will not be an assumption of the theory; it is used here to elucidate the possibilities. Even with this assumption, the game matrix (including the time components) might not be a constant. For each effective two-person game, the game matrix is constructed from the effective two-person game according to Eq. (1.8). The constant coalition flow for the composite space Eq. (1.9) defines a straight line curve in choice–space. I call two coalitions distinct if the corresponding partitions are distinct. I expect that in general, flows for distinct coalitions determine distinct lines. Thus in general, the time–choice components of Eq. (1.8) will be distinct for different coalitions. I expect therefore that the game matrix will not be constant due to the variation of the time–choice components. To reflect all the possible coalitions, the time–choice components of the game matrix interpolate between the specified values for each coalition. Thus, the static theory is incomplete. The normal attributes of static game theory deal primarily with the payoffs f_{mn} and not the time–choice components F_{m0}. I need to develop the dynamic theory further to gain insight into these new components.

I make a little progress in this direction. I consider a specific set of constant payoffs f_{mn} and a constant set of strategic flow ζ^m. An example might be one in which the strategies correspond to a coalition S. By considering the flow to be constant everywhere, in general I exclude the possibility that any other coalition has an equilibrium flow. This is sufficient to define a "fair" game for coalition S as an anti-symmetric matrix in the space of strategy–choices $\{x^m\}$ and time–choice $t = x^0$:

$$\begin{aligned} F_{mn} &= f_{mn} \\ F_{n0} &= \frac{1}{m_0}\zeta^m f_{mn} \end{aligned} \quad . \tag{2.1}$$

I call this the ***market field*** F_{ab} for the coalition S. I define the "stable" ***market flow*** V^a for coalition S to be the choice–space vector $V^a = \{\zeta^m \quad m_0\}$. From the formal rules it can be verified that the product of the market field and market flow vanish[16]: $F_{ab}V^b = 0$. For a two-person game, I recover Eq. (1.8).

If more than one coalition is active, I relax the definition above so that it holds only along the line defined by the coalition. If I can define an appropriate interpolation function to accomplish this, then I have a model that has more than one active coalition. I elaborate on this later in Chapter 6. For now, I take from this discussion that in general there will be a game matrix that is defined at every point of space and time, though it is one whose value may vary from position to position. The game matrix will specify the payout to each player, including the possibility for coalitions and factions at each point in choice–space. In general, the product of the market field and flow vector will not vanish.

2.3 Geodesics

I have articulated the rules-of-the-game at a high level but not the rules of the geometry in which these rules are embedded. As a first step towards articulating the geometry, I choose some basic conventions that will be useful. To help in distinguishing between strategy–choices and choices in the full choice–space, I define a convention that indices in the middle of the alphabet $m,n,\cdots$ represent strategy–choices only, whereas indices at the beginning $a,b,\cdots$ represent both strategy–choices and time–choice. Indices are mathematical trademarks that appear as subscripts as in the expression F_{ab}. As noted previously, some discussions are made more transparent by hiding the indices entirely such as $\mathbf{U}\cdot\mathbf{V}$. When there is minimal danger of confusion, I represent such quantities in bold, such as the market flow V^a as the vector $\mathbf{V}$ and the game matrix F_{ab} as the matrix $\mathbf{F}$.

I used the matrix notation provisionally to introduce the equivalency in Eq. (1.8) and Eq. (1.9). The transformation of the market flow $\mathbf{V}$ by the transformation matrix $\mathbf{U}$ is similarly introduced:

$$\mathbf{V} = \mathbf{U}\cdot\bar{\mathbf{V}}.$$

---

[16] Indices that are repeated between the product of two entities define an implied summation: $U_{ab}V^b \equiv \sum_b U_{ab}V^b$. The sign "$\equiv$" means "defined by".

The appropriate transformation of the game matrix requires two factors that are each inverse to the coordinate transformation matrix:

$$\mathbf{F} = \left(\mathbf{U}^{-1}\right)^T \cdot \overline{\mathbf{F}} \cdot \mathbf{U}^{-1} .$$

The notation is compact. Any particular game with N players is specified by the game matrix $\mathbf{F}$. Different points of view are related by a linear transformation. Implicit in this description is the underlying choice of strategies that can be chosen for any particular play.

To move to a dynamic theory of games, I require the notion of a distance and geometry. In the above, I have motivated the hypothesis that there is a market field and market flow defined in a choice–space such that provisionally they take on specified values along specified curves, such as along coalition flows, forming the boundary conditions but may take other values elsewhere. The question is "what values are in some sense ok?" I address this question by focusing on the invariance of the form of the equations in Chapter 4. Before embarking on a statement about the theory, I need more than just some notational help but some mathematical ideas from the field of geometry.

The origin of the word ***geometry*** is from two concepts: the earth and measuring. The early use of the word in English [Oxford English Dictionary (2002)] was in the practical art of measuring and planning. The study of geometry is much older. We may have learned about Pythagoras in school: The relationship between the sides of a right triangle is expressed as $c^2 = a^2 + b^2$. The relationship works on flat surfaces and is a good approximation if the surfaces are almost flat. Sailors have long used this approximation. On open seas, for the purposes of navigating between two points, the sea can be considered smooth. For relatively short distances, the sea can also be considered flat. Sailors determine their course by laying out rhumb lines on "flat" charts which cover only a portion of the sea. The collection of all the charts forms an ***atlas,*** which can be used to navigate around the world if the neighboring charts can be pieced together properly. The atlas provides a description of the geometry or "measuring on earth". It is easy to imagine that if the charts are made to cover areas that are small enough, the geometry will account for various approximations of what we think of as the earth, including it is neither flat nor smooth nor round. It can be as flat as the ancients believed, as long as the segment considered is not too large. It can be a round ball, which is pretty good for travel between distant points if we are not concerned with disturbances over water and the irregularities of land masses. For more accurate navigation it can recognize that the earth is more an oblate spheroid (squashed ball) than a sphere. Or in fact if we make our charts cover very small areas, it can account for waves, mountains and valleys, as well as houses, cars and people. Such an accounting would have to be done at each instant of time however since the shape of the earth at this level would be changing constantly.

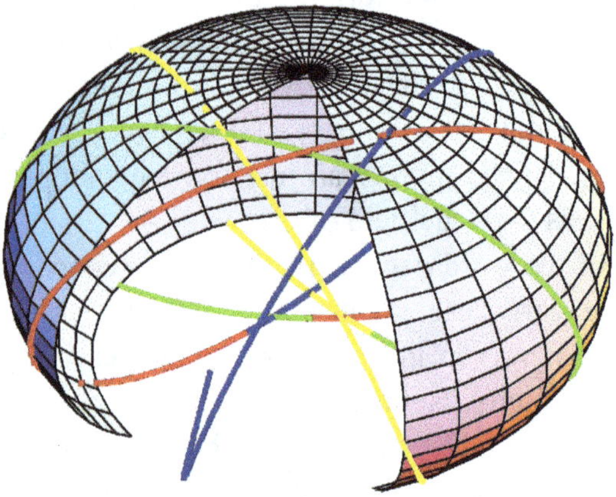

Fig. 2.1 Geodesics on an oblate spheroid; each color represents a successive continuation (blue-red-green-yellow) of the geodesic.

One of the most interesting problems in geometry and navigation is that of getting from one point to another as quickly and as accurately as possible. Such paths are called geodesics for obvious reasons. Navigators have dealt with this problem for thousands of years. The problem has gotten more complex however as we have come to a better understanding of the shape of the earth. As long as the earth was "flat", Pythagoras was good enough. The British Navy had to deal with the problem that the earth is a sphere and so officers had to learn their spherical trigonometry relations. On a sphere, the geodesics are closed great circles. In general however geodesics have complex global properties and are not in general closed but ***chaotic*** [*Cf.* Devaney (1989)]. Figure 2.1 would be a nightmare for such officers. Yet the answer is implicit in the formula from Pythagoras, as long as the distances are sufficiently small or "differential".

There is a great deal of power in the differential form of Pythagoras' formula:

$$dc^2 = da^2 + db^2 .$$

The differential notation, "d", indicates that small distances are considered. The form of the relation can be modified to take into account the units of measurement. A mariner can use nautical miles, while a land lubber can use feet, statute miles, meters, rods or whatever. If not all distances are in the same units, then a scale factor needs to be added to the relationship along each independent direction:

$$dc^2 = g_{aa}da^2 + g_{bb}db^2 .$$

There are other changes in the final form that might occur without changing anything fundamental about the actual length or distance measured.

We are familiar with measuring distances relative to directions that are orthogonal, such as using latitude and longitude for distances. Sailors are familiar with schemes where the distances are not orthogonal. You may know the angle to two different lighthouses. If these beacons (such as Loran) provide a radio signal, you may also know the distances as well. There is no particular reason for these distances to be orthogonal. Moving small distance away from each lighthouse will still result in a net distance as calculated by Pythagoras, by projecting the lighthouse distance along orthogonal coordinates. However, when measured relative to the two lighthouses, the form will involve an additional term:

$$dc^2 = g_{aa}da^2 + g_{bb}db^2 + 2g_{ab}dadb .$$

The extra term reflects the degree to which the two lighthouses do not provide an orthogonal system of measurements. This form in terms of a ***symmetric*** matrix $g_{ab} = g_{ba}$ is the most general: In other words, no matter what system of coordinates used, the "form" will be covariant: It will always look like the one above and produce the same value for distance.

The metric expression can be generalized to any number of dimensions, can include time and can allow for signatures other than Euclidean. To say that the metric expression always has the same "form" and produces the same value for distance is to say that the expression is covariant and describes a number which is invariant. A matrix notation is another way to display the invariance property:

$$ds^2 = \mathbf{dx}^T \cdot \mathbf{g} \cdot \mathbf{dx}.$$

The form is compact and makes visible the notion that a linear transformation of the coordinate distance $\mathbf{dx}$ to $\mathbf{dx} = \mathbf{U} \cdot \mathbf{d\overline{x}}$ will leave the distance invariant as long as the metric transforms appropriately: $\mathbf{g} = \left(\mathbf{U}^{-1}\right)^T \cdot \overline{\mathbf{g}} \cdot \mathbf{U}^{-1}$. The different possible geometries are now a statement about different possible symmetric metrics $\mathbf{g}$ that are not related by a linear transformation. We conclude that just as the game attributes are defined by the ***antisymmetric*** game matrix $\mathbf{F}$ defined by $F_{ab} = -F_{ba}$, the geometry is defined by the symmetric metric $\mathbf{g}$.

2.4 Games are Locally Flat

There is one additional subtlety. A metric is defined at each point in space and so the linear transformations under which the metric can be transformed can be different at different points: They are ***local***. This corresponds to looking at different charts in the atlas. It is a special attribute of certain spaces that they can be represented by charts which are locally flat and yet when stitched together produce spaces with complex curvatures. Moreover structures such as the game matrix are also defined at each point. Their variation from point to point is constrained by the rules of differential geometry by the same requirement that they match when the charts are stitched together.

The subject of differential geometry is the study of such spaces with symmetric metric structures as well as other "tensor" structures defined on a complete atlas in any arbitrary dimension. Any symmetric matrix can be transformed with a linear transformation into a diagonal matrix consisting of the numbers ± 1. The metric corresponds to a specific type of symmetric matrix that transforms to the identity matrix: All components are $+1$. By considering any symmetric matrix, differential geometry allows a generalization to Pythagoras' geometry, called Euclidean geometry. A way to characterize the new possibilities [E.g. Synge and Schild (1949)] is with the ***signature*** of the symmetric matrix, defined as the difference between the number of "plus ones" and "minus ones". Differential geometry requires that a metric have the same signature at every point in order that different neighboring charts match up properly. The metric and its signature therefore determine the local geometry.

I use the language of differential geometry to articulate the dynamic theory of games. I require not only a symmetric metric but other types of structures on the space. These can be introduced by first discussing what appears to be a notational issue. I recall that compact notation makes general results obvious but without adequate machinery, makes specific results hard to fathom. Conversely, less compact notation often make calculations relatively easy but may obscure the general result. As in the past, the compromise is to do a little of each. I have already introduced both the matrix notation and the index notation. I use the matrix notation to indicate certain general results but use a "component" based notation to make calculations appear more plausible. The index notation also helps to articulate the possible structures that can be built on the space.

The component vector $\mathbf{dx}$ becomes dx^a. The index spans the time–choice and the possible player strategy–choices. However, I need one new distinction that is dependent on how a quantity transforms or changes under a linear transformation. In discussing the transformation properties, I have two possibilities: Quantities that transform according to the coordinates and the matrix $\mathbf{U}$ and quantities that transform according to the inverse of this matrix, $\mathbf{U}^{-1}$.

As an example, pressure p over the earth is thought of as a ***scalar*** function, having value but no direction: Its value is the same for all observers whether stationary or moving. Those who study weather know that by looking at isobars, properties of the weather can be deduced[17]. Gradients or changes in the pressure cause air to flow. These changes depend on direction, so the incremental change in pressure is a linear function of the incremental changes in distance:

$$dp = p_{;a} dx^a .$$

The sum over the various components indicates the linearity and the notation for the coefficients emphasize that they are related to pressure. The "semi-colon" before the index is a new short-hand that the relationship is differential: The pressure gradient $p_{;a}$ is defined as the variation of the pressure along the direction dx^a. The total change in pressure does not depend on the specific coordinate system and so its value and form remain unchanged under a linear transformation:

$$dp = p_{;a} dx^a = \overline{p}_{;c} \left(U^{-1}\right)^c{}_a U^a{}_b d\overline{x}^b = \overline{p}_{;c} \delta^c_b d\overline{x}^b = \overline{p}_{;b} d\overline{x}^b .$$

The "Kronecker delta" matrix δ^c_b is a diagonal unit matrix. The result shows that with an appropriate transformation, the pressure differential does not change. The resultant form suggests a distinction that is important: Quantities that transform like a coordinate dx^a (*i.e.* they transform with $\mathbf{U}$) will be indicated with an ***upper*** index. In differential geometry, the index is termed contra-variant. Quantities that transform like $p_{;a}$ (*i.e.* they transform with $\mathbf{U}^{-1}$) will be indicated with a ***lower*** index. In differential geometry, the index is termed co-variant. If a sum occurs over an upper and lower index, it is called contracted and the result (remember the implied summation) is always invariant as far as those two indices are concerned because of their opposite transformation properties. To make the result transparent in more complicated expressions, there is a slight modification to the summation convention: Unless stated differently, the implied summation will be restricted to an upper index repeated by a lower index.

---

[17] I believe the weather provides a useful analogy. There is a readable book on weather for sailors by Crawford (1992) that I have found quite entertaining and instructive. Modern weather was constructed by Vilhelm Bjerknes. A readable history has been done by Friedman (1989). The origin of the terms "warm front" and "cold front" becomes quite clear after reading this.

An application of this rule is that the line element in terms of the metric is $ds^2 = g_{ab}dx^a dx^b$. This transforms to the following under a linear transformation:

$$ds^2 = \overline{g}_{ef}\left(U^{-1}\right)^e_{\ a}\left(U^{-1}\right)^f_{\ b} U^a_{\ c} U^b_{\ d}\, d\overline{x}^c\, d\overline{x}^d = \overline{g}_{cd}\, d\overline{x}^c\, d\overline{x}^d .$$

This is the "proof" or restatement of the definition that the distance is an invariant. I also make connection with the matrix notation:

$$\mathbf{g} = \left(\mathbf{U}^{-1}\right)^T \cdot \overline{\mathbf{g}} \cdot \mathbf{U}^{-1} .$$

This form can also be used to express the fact that the distance is an invariant.

The conclusion is that the geometry is locally flat and has in addition to a symmetric metric structure and an "anti-symmetric" game matrix structure, other structures that can be classified as having some number of upper and some number of lower indices. The upper and lower indices indicate their transformation properties. It will be the purpose of the theory to indicate which structures are necessary and provide rules for calculating all such structures that are part of the theory.

2.5 Dynamic Game Theory Hypothesis

I summarize the ***Dynamic Game Theory Hypothesis***: There is a choice−space composed of time–choice and one dimension for each player (pure) strategy–choice on which exists (from the rules-of-the-game) an antisymmetric game matrix F_{ab} and a market flow $V^a = dx^a/ds$ and a symmetric metric field g_{ab} (the new ingredient). These components reflect a local behavior. The dynamic aspects are determined by the symmetric metric g_{ab}, which provides the notion of a length or line element $ds^2 = g_{ab}dx^a dx^b$ whose value and form are invariant under local linear transformations. The dynamic theory makes contact with the static theory at certain special points or boundaries ζ^a in the ***null space***[18] of the market field: $F_{ab}\zeta^b = 0$.

---

[18] The null space is a special distinction in geometry and group theory [E.g. see the elementary text Birkhoff and MacLane (1959)] based on the observation that the set of all vectors Z^b whose products satisfy $F_{ab}Z^b = 0$ form a subspace of the full domain. The

The Game Theory Hypothesis defines the theory of games as a geometric construct. The flow is normally termed a ***vector field*** and the game matrix and metric are examples of ***tensor fields***. The mathematical distinctions highlight the transformation properties that I have articulated above. Such geometric structures are well understood and described in the literature. Therefore we appear to have a solid structure from which to proceed. The next chapter takes up the question of what are reasonable forms for the geometry.

---

subspace so defined is independent of the reference frame used to define it. As such it is an insightful and invariant attribute determined by the matrix F_{ab}.

Chapter 3

Flow of Strategic-Mass

The goal of this chapter is to be more specific about the dynamics of games and how the dynamics follows from geometry. The geometry provides the framework in which games move. There is simplicity in focusing on the metric, market field and flow since they directly relate to the assumptions about measurements and payoffs. One complication is that they are local attributes, not global attributes. A second complication is that an understanding of the metric may lead to an understanding of motion only if no "external" forces are present. Both complications will be addressed here. In particular, I extend a Newtonian concept of motion to game theory: **The flow of strategic-mass is constant unless acted upon by external forces.**

3.1 Local versus Global

To understand the ideas of local and global attributes, I return to a familiar example: the geometry of the earth. We think of the earth as a spherical ball. That is a clearly defined and commonly understood vision. We navigate over the earth measuring distances in terms of the longitude ϕ and latitude θ. To allow for the possibility that the earth is not a sphere, we require knowledge of the variation of the radius r in terms of the longitude and latitude. These quantities determine the relationship of distance by means of the differential distance or ***line element***[19].

[19]The line element depends only on the latitude and longitude since the radius is a known function of these variables: $dr = r_{;\theta} d\theta + r_{;\phi} d\phi$. For a sphere, this vanishes; for an oblate spheroid, the radius is a function of latitude only, $dr = r'(\theta) d\theta$.

The line element is useful for navigation, pointing out that only distances along the equator are given directly in terms of the longitude:

$$ds^2 = r^2 d\theta^2 + r^2 \sin^2 \theta d\phi^2 + dr^2 . \tag{3.1}$$

The challenge for the differential geometry formulation is to make the "big picture" visible. In particular, the above line element does not appear to show which paths are the shortest (or longest) over the surface. Yet most of us believe we are familiar with those paths: They are great circles. We might experience this directly if for example we travel by air between two cities on a great circle route.

For more complicated line elements such as implied by Fig. 2.1, the geodesics are not well known, even by sailors. However I assert that the above equation for the line element is enough. The challenge is to show this is at least plausible. The starting point is the general line element $ds^2 = g_{ab} dx^a dx^b$ between two neighboring points separated by a differential distance dx^a. This defines the distance locally. A path between two distant points will have a length that is constructed by adding up these local distances. Now I do a thought experiment: I imagine doing this for all possible paths. Of this set of paths, there will be a subset of paths that are shorter than all the rest. A typical property of something being "shortest" is that nothing nearby will be shorter.

This idea can be sharpened by comparison to another "familiar" idea. Consider Fig. 3.1. It shows a river that runs down into a valley floor and then into a mountain lake. Over time, we know that the stream finds and gouges out an optimal path as it runs down into the valley, a fact that accounts for the meandering path observed. What is this optimal path? Water flows downhill unless it finds a barrier, a dam. Once the dam overflows, the water continues its path downhill. At some point, the dam might not overflow and a lake is formed. The water knows locally the difference between downhill and uphill; unblocked it never flows up. The optimal path the water takes and the shortest path discussed in the previous paragraph have a lot in common. I use mathematics to summarize the essential content. Near a minimum, the form of the height versus the horizontal distance has the form of a parabola: $y = x^2$. The variation of height $\delta y = 2x\delta x$ vanishes at the origin independent of the variation δx. This is summarized by the condition $\delta y = 0$.

Fig. 3.1 Mountain Stream in the Vaudois Alps was photographed by the author.

Returning to our variation problem for the geodesic, the set of all possible paths is like an infinite collection of little parabolas corresponding to varying the position of the path at every point and so a compact statement of conditions for the path to be the shortest is that the total length be at the collective set of "shortest positions":

$$\delta \int ds = 0 .$$

This compact notation represents the thought experiment and so represents a lot of work. However, there exists mathematical machinery that summarizes the common approach to such problems.

So what is the machinery for constructing a minimal path? Two paths may differ at every point, so a label τ is needed to describe the path differences at each point:

$$\delta \int ds = \delta \int \frac{ds}{d\tau} d\tau = \delta \int \sqrt{g_{ab} \frac{dx^a}{d\tau} \frac{dx^b}{d\tau}} d\tau = 0 .$$

At each point labeled by τ, the incremental length depends on position x^a and the relative change in position along the path or "velocity" $\dot{x}^a = dx^a / d\tau$. The incremental change in length or speed depends on:

$$L\left(x^a, \frac{dx^a}{d\tau} \right) = \sqrt{g_{ab} \frac{dx^a}{d\tau} \frac{dx^b}{d\tau}} .$$

These problems occur in mathematical physics frequently and are called variation problems[20]. They have a general solution:

$$\frac{d}{d\tau}\left(\frac{\partial L}{\partial \dot{x}^a} \right) - \frac{\partial L}{\partial x^a} = 0 . \tag{3.2}$$

This solution results from carrying out the above mentioned thought experiment:

$$\delta \int L d\tau = \int \left(\frac{\delta L}{\delta x^a} \delta x^a + \frac{\delta L}{\delta \dot{x}^a} \delta \dot{x}^a \right) = \int \left(\frac{\delta L}{\delta x^a} \delta x^a - \left(\frac{d}{d\tau} \frac{\delta L}{\delta \dot{x}^a} \right) \delta x^a \right) .$$

Two neighboring paths will differ not only by position but by velocity. There will be a contribution from each in computing the total change in length. The change in velocity is proportional to the change in position. Integration by parts provides the proportionality constant. The solution to the variation problem demands that the above variation vanish at every point, since the variations at each point are independent of each other. Equation (3.2) is the standard way of stating the result.

Equation (3.2) is powerful. It provides the shortest distance between two points entirely in terms of the metric and its derivatives. It provides

[20] A physical way of seeing this is Geometrical Optics, which is provided in volume 1 of Feynman (1963).

the answer to the question posed at the outset: It gives the big picture in terms of the line element. For reference, I summarize the result of the computation for the geodesic:

$$\begin{gathered} g_{ab}\frac{d\dot{x}^b}{d\tau}+[a,bc]\dot{x}^b\dot{x}^c=0 \\ [c,ab]\equiv \tfrac{1}{2}\left(-\partial_c g_{ab}+\partial_a g_{bc}+\partial_b g_{ca}\right) \end{gathered} \quad . \tag{3.3}$$

These equations define the geodesic that goes between two points in an arbitrary geometry in terms of the metric and, using the last equation, a ***bracket*** that is defined as the shown linear combination of gradients of the metric. I make the following observations about this equation:

- First I note that it is a second order differential equation for the coordinates, which puts it into the class of problems that are easily solved numerically. That is useful and practical. A familiar example of a first order equation is the classic growth problem of a population. The population grows at a rate proportional to the size of the population. A more realistic problem generates a second order differential equation, which takes into account the decline of the population due to for example, a predator. The death rate could be proportional to the number of predators. Moreover the number of predators grows at a rate proportional to both its population size, as well as the population size of its food, the initial population. The two first order equations are coupled, generating a second order differential equation. System Dynamics provides techniques and software [*Cf.* High Performance Systems (1997)] for solving such coupled problems.
- Second, I note that there are terms that depend on the variation of the metric along x^c: $\partial_c g_{ab}=[a,bc]+[b,ca]$. Such terms are often termed *fictitious forces*; the Coriolis force is an example. It is an artifact of the observer being in a rotating frame. An observer in space observing the rotation of the earth would not require such forces. Remember that the concept of a metric allows different observers to view the same geometry and in particular makes no restriction on the motion of such observers. Such fictitious forces arise from the transformation properties between different observers.

- Third, I note that it is always possible to find a local coordinate system in which the metric is locally flat. This is the case for navigating on the surface of the earth using "flat" charts. Technically, this is the statement that the metric can be chosen to be Pythagorean [diagonal] and minimal [its first derivatives vanish]. In such a coordinate system, the geodesics have constant velocities: The acceleration *is zero.* In other words, the motion corresponds to a **flow in which there are no external forces**. In the next section I show this is true independent of observer.
- Fourth, the non-trivial or "big picture" aspects of geometry appear with the second derivatives of the metric. As in the variation problems, the curvature makes its appearance with the second derivatives and so does not contradict the local point of view that states that the surface is approximately flat (to first order in the derivatives).

All of this is consistent with the statements made earlier about the geometry and its determination by local charts.

3.2 The Connection

A main goal of differential geometry is to isolate the *forms* of geometric structures [mathematical trade-marks or objects] that do not depend on coordinate system. These structures characterize geometry. The shortest path should be one such structure. The geodesics depend on the derivatives of the metric. To understand how the form of the equation depends on a coordinate system, I have to say a little bit about the metric, its derivatives and recall their transformation properties that were defined in the last chapter.

The metric transforms as a ***"tensor field"*** with two lower indices, that is as a structure defined at each point in space. I indicated in matrix notation its transformation properties: $\mathbf{g}=\left(\mathbf{U}^{-1}\right)^{T}\overline{\mathbf{g}}\mathbf{U}^{-1}$. The metric is a non-singular matrix and so has an inverse whose transformation properties follow: $\mathbf{g}^{-1}=\mathbf{U}\overline{\mathbf{g}}^{-1}\mathbf{U}^{T}$. This shows that the inverse transforms like a tensor field with two upper indices. The inverse is written using the same root label with upper indices: g^{ab}. The use of the same root label highlights the result from differential geometry that the two tensors represent the same geometric structure. In differential geometry tensors, that include scalars and vectors, are structures that transform linearly according to their upper or lower indices: Scalars have no indices, vectors have one index and ***tensors*** are the generic term and can have any number of indices. Tensors are structures defined at a point; the term ***tensor field*** is used if the tensor is defined at every point in space.

Given a tensor field, other tensors can be constructed. For example, any tensor with an upper index $T^{\cdots a \cdots}_{\cdots}$ can be converted to a tensor with a lower index by multiplication and contraction with the metric: $g_{ab}T^{\cdots b \cdots}_{\cdots}$. The new tensor is given the same name since it describes the same geometric structure $\mathbf{T}$. The same conversion can be done with any lower index, multiplying and contracting with the inverse metric. Another example is the product of two tensors such as $U^a V^b$, which is again a tensor.

Not all operations however yield new tensors. Specifically, there are problems when comparing the values of tensors at different points. For example, the difference of the numerical value of the metric g_{ab} between two points separated by the differential distance dx^c is represented by $dx^c \partial_c g_{ab}$. However, the factor $\partial_c g_{ab}$ does not obey a simple transformation rule. The transformed factor depends not just linearly on the transformation matrix $\mathbf{U}$ but on the transformation matrix derivative $\partial_a \mathbf{U}$. This is true of the bracket quantity in Eq. (3.3). Does this mean that the *form* of the geodesic is not the same in different coordinate systems?

The problem is traced in differential geometry to how one makes comparisons in computing derivatives. For a scalar function, there is no problem. The derivative of a scalar p relative to a coordinate x^a at a point is computed by first considering a curve that goes through the point in the direction x^a. Evaluate the function at the given point and along the curve a small distance away, take the difference of the two functions and divide by the small distance. No additional dependence on the transformation matrix $\mathbf{U}$ is required. The limit of an infinitesimal distance provides the variation $p_{;a} = \partial_a p$. The notation $p_{;a}$ is reserved for variations that are tensors; the notation $\partial_a p$ is reserved for the standard method used above for taking derivatives. In this case they are equal.

Differential geometry provides a rule for comparing the values of tensor objects at different points. I have already suggested what that rule is: Pick a special frame in which the space is locally flat and make the comparison. Though correct, it seems to make one coordinate system more special than all the rest. The solution of expressing the same idea irrespective of coordinate system requires more mathematical distinctions. In particular, differential geometry introduces the notion of ***parallel transport*** of geometric [tensor] structures. The motion of a tensor structure along a geodesic path is linear in the tensor. For an infinitesimal movement, the following summarize the rules for the ***covariant differential*** change in a tensor in terms of the proportionality constant Γ^c_{ab} called the ***connection***:

$$X_{a;b} = \partial_b X_a - \Gamma^c_{ab} X_c$$

$$X_{ab;c} = \partial_c X_{ab} - \Gamma^d_{bc} X_{ad} - \Gamma^d_{ac} X_{db}$$

$$\ldots$$

$$X^a{}_{;b} = \partial_b X^a + \Gamma^a_{bc} X^c$$

$$X^{ab}{}_{;c} = \partial_c X^{ab} + \Gamma^a_{cd} X^{db} + \Gamma^b_{cd} X^{ad}$$

$$\ldots$$

The connection is a ***messenger*** potential that communicates the space-time structure from one point to another. Göckeler and Schüker (1987) show that this is analogous to the vector potential of electro-magnetism and thus demonstrates that differential geometry with a connection is a ***gauge field theory***.

Our usual notions of geometry are summarized by the condition that the parallel transport of the metric yields numerically the same metric. This implies the covariant differential change is zero:

$$g_{ab;c} = \partial_c g_{ab} - \Gamma^d_{bc} g_{ad} - \Gamma^d_{ac} g_{db} = 0\,.$$

A straightforward application of the mathematical rules shows that these conditions determine the connection in terms of the metric:

$$\Gamma^c_{ab} = \tfrac{1}{2} g^{cd} \left(-\partial_d g_{ab} + \partial_a g_{bd} + \partial_b g_{da} \right).$$

This explicit expression allows the transformation rules for the connection to be computed and from them prove that the linearly transported structures defined above $X^a{}_{;b}$ *etc.* indeed transform as tensors. This is despite the fact that the connection does not transform as a tensor. Indeed, the connection is directly related to the bracket by Eq. (3.3) that defines the geodesic: $\Gamma^c_{ab} = g^{cd}\left[d, ab\right]$. Furthermore, in a system that is locally flat, the connection vanishes[21]. This extends the rule that comparisons are made in a locally flat system. The connection allows differential statements (that compare tensors at neighboring points) to be made that are valid in all coordinate systems. As a first example, in any coordinate system, the equation for a geodesic is:

$$\dot{x}^a_{;b}\dot{x}^b = 0 .$$

The form looks slightly different because I recognize that the velocity is a vector field and its variation along the path can be computed by considering the "chain rule": The full co-variant variation consists of summing the product of the variation along each coordinate direction with the variation along that coordinate direction:

$$\left(\dot{x}^a_{;b}\right)\dot{x}^b .$$

The expression follows by using the form for the parallel transport $\dot{x}^a_{;b} = \partial_b \dot{x}^a + \Gamma^a_{bc}\dot{x}^c$ and the definitions.

Looking at the rules I reiterate that neither the connection nor the ordinary derivative separately have transformation properties of a tensor. The full expressions however are tensors. To distinguish quantities such as $X^a{}_{;b}$ from ordinary derivatives they are called ***co-variant derivatives***. Since locally, there is a coordinate system in which the connection terms vanish, the co-variant derivatives are extensions of ordinary derivatives. The co-variant derivatives provide a covariant generalization to the chain rule:

$$\frac{DX^a}{\partial\tau} \equiv X^a{}_{;b}\dot{x}^b .$$

---

[21] By the way, it is the connection that expresses such fictitious forces such as the Coriolis force. The term "fictitious" is justified because the forces are artifacts of a particular choice of coordinate system. They are not attributes of the invariants of the geometry.

In terms of this chain rule, the equation for the geodesic takes on the compact form:

$$\frac{D\dot{x}^a}{\partial \tau} = 0 \,.$$

I now can answer the question about the geodesic and whether the form of the equation is the same in different coordinate systems. It is the same because it can be expressed using only covariant derivatives:

$$\frac{D\dot{x}^a}{\partial \tau} = \dot{x}^a_{;b} \frac{dx^b}{d\tau} = \left(\partial_b \dot{x}^a + \Gamma^a_{cb} \dot{x}^c\right) \dot{x}^b = \frac{d\dot{x}^a}{d\tau} + \Gamma^a_{bc} \dot{x}^b \dot{x}^c = 0 \,. \qquad (3.4)$$

The geodesic is defined by the vanishing of the covariant acceleration. The expression is a tensor relation and valid in any frame. The motion is along a geodesic if there are no external forces, using the Newtonian concept that forces cause acceleration. The geodesics in the last chapter, Fig. 2.1, were computed using the line element for an oblate spheroid.

3.3 Curvature

I maintain space is locally flat. What does space look like on a larger scale and what geometric structures determine this large scale behavior? The familiar example of the earth suggests that structures that reflect the curvature also reflect the large scale behavior. For example, the geodesics seem to reflect the curvature of space. Since the geodesics depend on the second derivatives of the metric, it is reasonable to expect that all curvature characteristics might be so determined. I now show that this expectation is met.

Differential geometry pursues the notion of parallel transport not just along a small segment of a curve but around small closed curves. Consider as an example a very small parallelepiped and start with two adjacent sides. Each side defines a very short curve in space and that curve defines a direction and hence a vector field. There are two such vector fields corresponding to the two directions, X^a and Y^a, along each short curve. The trick is to consider now a third vector Z^a defined where the sides meet and transport it to the point diagonally opposite on the parallelepiped. There are two ways to get there: I can transport first along X^a and then along Y^a or first along Y^a and then along X^a. In either case, the answer should be proportional to each of the vectors and so the difference is also proportional: $R^a{}_{bcd}X^bY^cZ^d$. Since parallel transport generates a vector, the proportionality factor $R^a{}_{bcd}$ is a tensor. If space were flat everywhere, not just locally, this tensor would vanish: It wouldn 't matter which path you took to get to the opposite diagonal. On a curved surface however, it is not hard to be convinced that the result is not zero. The trade-mark $R^a{}_{bcd}Y^cZ^d$ is analogous to the electro-magnetic ***messenger*** field passing through a surface [Göckeler and Schüker (1987)]; flat space corresponds to no metric messenger field.

For this reason the object $R^a{}_{bcd}$ is called the ***curvature tensor***. Since the parallel transport along each leg is determined by the connection, it should not be a surprise that the curvature tensor is determined by the connection (or equivalently the "bracket") and its derivatives:

$$R_{abcd} = \begin{Bmatrix} \tfrac{1}{2}\left(\partial_b\partial_c g_{ad} + \partial_a\partial_d g_{bc} - \partial_b\partial_d g_{ac} - \partial_a\partial_c g_{bd}\right) \\ + g^{ef}\left([e,ad][f,bc] - [e,ac][f,bd]\right) \end{Bmatrix}.$$

The explicit expression shows clearly that even locally when the first derivates of the metric vanish, the curvature tensor does not. In general the curvature tensor depends on the first and second derivatives of the metric.

From the full tensor, a second order symmetric tensor ("matrix") $R_{ab} = R_{cabd}g^{cd}$ can be created by contraction and its form simplified in terms of the metric and determinant $g \equiv \det \mathbf{g}$:

$$R_{ab} = \partial_a\partial_b \ln\sqrt{|g|} - \partial_c\Gamma^c_{ab} - \Gamma^c_{ab}\partial_c \ln\sqrt{|g|} + \Gamma^d_{ac}\Gamma^c_{bd}. \tag{3.5}$$

Differential geometry maintains that the big picture properties of space are determined by these curvature tensors. The symmetric tensor will be an important key to understanding the theory as set by the geometry, as well as the further contraction $R = g^{ab}R_{ab}$ that defines the curvature scalar. The symmetric curvature tensor satisfies the set of identities:

$$R_{ab;c}g^{bc} = \tfrac{1}{2}R_{,a}.$$

This set of identities is satisfied independent of the choice of metric. I introduce the following combination of curvature tensors[22] called the ***energy–momentum tensor***:

$$-\kappa T_{ab} \equiv R_{ab} - \tfrac{1}{2} g_{ab} R . \qquad (3.6)$$

In analogy with physics I say that this tensor is ***conserved*** if:

$$T_{ab;c} g^{bc} = 0 .$$

This result in fact follows from the above mentioned identities satisfied by the curvature tensor. In particular it is a distinction that is purely geometric in origin. [Göckeler and Schüker (1987)]. Different possible spaces are distinguished by their having different conserved energy–momentum tensors.

3.4 Geometry Specified by Sources

The conservation properties of Eq. (3.6) led Einstein to inquire whether this tensor might not be an extension to the physical energy momentum tensor constructed in mechanics. It was a shift in perspective. In the example of the earth, I focused in the last section on its shape and how that shape can be made visible from the line element. It is plausible that the large scale nature of the shape is reflected by the curvature and so many properties of that shape are contained in the tensor T_{ab}. In the last section I also focused on a special set of motions: Those motions that occur without any external force and are characterized as geodesics.

Within the context of any given geometry there can be external forces that cause the motion to be different from a geodesic. The motion of the atmosphere around earth dictates the weather [Crawford (1992)]. Acceleration occurs because of these external forces. In physics, acceleration determines energy and momentum, defining an energy momentum tensor. In particular a fluid, because of its mass, will generate a gravitational force that pulls it together. At the same time, a fluid is forced apart because of its pressure. Our atmosphere would disappear because of this pressure if it were not for the gravitational force. Planets

---

[22] The factor κ is an arbitrary numerical constant of the theory.

such as Jupiter which are presumed to be entirely fluid are even more dramatic examples of this phenomenon. How much of the motion of such fluids is geometry and how much of the motion is determined by external forces?

The leap that Einstein made was a particular decomposition of the forces into those that are reflected in geometry and those that remain external. That leap provided an understanding of the shapes of planets and stars that was an extension of the results of Newton. It suggests an approach that might also work for the dynamic theory of games. The basic idea can be phrased as follows: The geometry contains long range effects such as gravity whereas the short range forces are expressed as external forces. Short range forces are typical properties of fluids and have properties such as density, pressure, elasticity, viscosity, *etc.* The nature of the short range forces can be more empirical and initially does not require a detailed understanding of their origin. Therefore the form taken by such forces will be based on considerations of simplicity and symmetry. Finally, the geometric energy momentum tensor T_{ab} is identified with the energy momentum of the short range forces. I say that these short range forces are generated by an elastic medium, the external strategic sources. An example of the energy momentum tensor generated by such forces is Eq. (1.4).

Since the theory of games needs to stand alone, I provide game–theory justifications for these external ***strategic sources***. First, I take it as an axiom that the acceleration of flow is caused by external sources. I assert that a perfectly good place to start is to relate T_{ab} to some set of external strategic sources. The real question is to understand the origin and properties of such sources. One such source must be the result of the static theory of games. In addition, I imagine that in a sequence of plays, neighboring plays are related to each other by some elasticity. They are not independent events but bound together like an elastic fluid forming a streamline through space. The subtle change is that neighbors that are not part of the sequence are also bound together by the same elastic forces. [I note that thermodynamics for fluids can be constructed that is consistent with the game theory proposed here (*Cf.* Appendix A).]

This may sound like a reach. Yet, it is true that in economics the word "elasticity" is used to describe supply–demand curves. Elasticity is ascribed to a market and suggests that independent plays are in fact related to each other. The players in economic games communicate with each other to some degree that is not captured by normal game theory. Von Neumann and Morgenstern (1944) used such arguments in creating the concept of stable sets of behaviors, where coalitions rise and fall. That certain behaviors die out and others survive is a Darwinian notion called ***structural coupling*** by Winograd and Flores (1986). For this to occur there must be some communication, some elasticity.

I imagine multiple games going on at any point in time, like a market situation. Each game is part of a sequence of games, has an identity like a company in a market. The idea of elasticity must play a role along the streamline of any one game and between simultaneous and neighboring games. The medium summarizes their collected wisdom, their interconnectedness. It is the medium that generates the geometry. It is market wisdom that determines whether the motion along a streamline is sluggish or volatile. Over time, it is the collection of games that operate near the fixed point that creates a type of gravity, attraction or structural coupling for rational behavior. Behaviors that are irrational still move in this gravity field and their motion reflects how far out they are from "center". There should be penalties as well for very irrational behaviors.

Based on these heuristic arguments, I articulate the foundations of the theory. The metric and, in the absence of external forces, the path a game takes from one point to another is determined by the geodesic. The metric is determined by the external sources by means of the energy momentum tensor, which determines the curvature of space. As a start, I take the sources to be a fluid, Eq. (1.4). I next deal with the market field that determines the static game theory behavior.

As suggested in the Introduction, there is an elegant way to introduce this interaction: ***Value–Choice Hypothesis***. Not long after Einstein introduced his theory of gravity, that included in it a theory of electricity and magnetism describing a charged fluid, researchers [Kaluza and Klein][23] noticed that this theory of two long range forces could be unified into a single theory by adding one new but hidden dimension of space.

[23] An early review of the Kaluza and Klein work is in Supplemental Note 23 in Pauli (1958).

The modern concept of string theory, with a much richer understanding of the short range forces, generalizes this observation. The relevance to the theory of games is based on a second result first observed by Von Neumann and summarized by Luce and Raiffa (1957) that a possible form for computing game theory results could be based on the form of Eq. (1.10) with only the first term. This is the equation for a charged fluid under the influence of an electro–magnetic field in strategy–choice space. I am led to a provisional proposal to describe the market forces and the external medium for the strategic sources that generalizes in a simple way the above foundations for the theory.

I consider an elastic fluid in choice–space with additional dimensions. I introduce a hidden value–choice coordinate ξ^j for each player j. I assume that the metric, market potential and market flow have no dependency on this value–choice. To distinguish this ***expanded choice–space*** from the ones considered up till now, I use the notation $x^\mu = \{\xi^j \quad x^a\}$; Greek indices indicate I am considering this expanded space.

In this monograph, unless otherwise noted, I make the additional ***Stationary Hypothesis*** that the metric, market potential and flows are independent of time. I will show later [*Cf.* Appendix C] that the time–choice can be viewed as a hidden coordinate $\xi^t = x^0$ in the space of strategy–choices, so that gravity effects become part of the general structural coupling γ^{jk} in Eq. (1.10). In addition however there are contributions that appear like a game payoff matrix. This additional complexity provides the possibility that time–choice might dictate a game different from a player's perception.

The Value–Choice Hypothesis implicitly assumes that the metric $\hat{\gamma}_{\mu\nu}$ is determined solely by the fluid flow in the expanded choice–space:

$$R_{\mu\nu} - \tfrac{1}{2}\hat{\gamma}_{\mu\nu}R \equiv -\kappa\left((\mu + p)V_\mu V_\nu - p\hat{\gamma}_{\mu\nu}\right). \qquad (3.7)$$

In the next chapter I outline the steps needed to express this hypothesis in terms of the strategic metric g_{ab} and the market fields F^j_{ab} [*Cf.* Eq. (4.4)]. A consequence of this hypothesis is that the market field is a source for the strategic metric even in the absence of any other strategic sources. I also obtain the motion based on the combined set of market and elastic forces, Eq. (1.10). I see explicitly the contribution to the acceleration from the market field and the strategic source, the fluid. If I further assume the Stationary Hypothesis, the same equation extended to include time–choice as a hidden coordinate shows that gravitational forces are part of the structural coupling γ^{jk}. In addition, there are forces that look like market forces $V_0 F_{ab}^0 V^b$ associated with the time–choice. So, not only does it matter how each player sees the game, it also matters how the game is perceived over time. It is as if time acts like an impartial player.

I get preliminary insight into the origin of the acceleration of strategic-mass in the expanded choice–space. The fluid flow can be taken to be described by a time-like unit vector, $\hat{\gamma}_{\mu\nu} V^\mu V^\nu = 1$. Such flows are called time-like if there is a frame in which the metric is diagonal and the normalization equation has the form $\left(V^0\right)^2 - \left(V^1\right)^2 - \cdots = 1$ showing that the time component is larger than the sum of the space coordinates[24]. Indeed, it can be shown that there is a frame in which the space coordinates vanish and the flow has only a time component. This is a useful property when performing calculations.

Because the energy momentum tensor is conserved, the functions that describe the fluid are determined by Eq. (3.7):

$$\begin{aligned} &V^\lambda \rho_{;\lambda} + \rho V^\lambda_{;\lambda} = 0 \\ &V^\lambda \mu_{;\lambda} + (\mu + p) V^\lambda_{;\lambda} = 0 \\ &\dot{V}^\mu = V^\mu{}_{;\lambda} V^\lambda = \left(\hat{\gamma}^{\mu\lambda} - V^\mu V^\lambda\right) \frac{\partial_\lambda p}{\mu + p} \end{aligned} \qquad . \qquad (3.8)$$

I have the basic result for this chapter, namely the equations that describe the flow of the strategic-mass.

However, the form of the equations does not yet distinguish the payoff matrix from the gravitational effects. This will be done in the next chapter. Here I note some of the basic properties. The first equation is the defining equation for the conservation of the market density. Market stuff or strategic-mass moves in such a way that market density or strategic-mass is neither created nor destroyed. The flow of market density into or out of the cell is accounted for by the increase or decrease

[24] If the sum is negative, the vector is called space-like; if the sum is zero, the vector is called a null-vector.

of the stuff in the cell. The next equation follows from the conservation of energy momentum and is a statement that the change in internal energy is determined by the "mechanical" property of the medium being compressed. The third equation, also a consequence of the energy momentum conservation, states that the acceleration of the flow is given by the transverse gradient of the pressure.

As a further help or insight to these equations, I note a more familiar yet still quite complicated example of a weather map, which is based on the same ideas: A weather map describes air, a fluid in motion, characterized by pressure isobars and allows us to infer the motion of the wind, the wind moving from high pressure to low pressure, bending to the right in the Northern hemisphere because of the rotation of the earth (Coriolis force). The last equation has both concepts: The right hand side describes the forces acting on the wind to move from high pressure to low pressure. The left-hand side describes the acceleration as seen from any frame of reference and in particular will produce the fictitious (Coriolis) force due to the motion of the reference frame (the observer sitting on the moving earth).

A consequence of this thinking is a focus on the flow of the medium, as described by a flow vector V^μ. It has a value over the entire space–time of the geometry. It describes the average flow at any point in space–time. Because this is defined at every point in space–time, the vector field will define an ***integral curve*** through any given initial point relative to a parameterization s:

$$\frac{dx^\mu(s)}{ds} = V^\mu\left(x^\mu(s)\right). \tag{3.9}$$

These equations are coupled first order differential equations and provide a unique curve through the given initial point. The solution, called a ***streamline***, will not in general be a geodesic since the acceleration does not vanish. The motion is determined by Eq. (3.8). The acceleration is seen to be proportional to the transverse pressure gradient.

The flow may change over time or remain stationary, even though by definition the flow represents motion of the fluid. Think of a gentle breeze blowing from the South. The flow of air around you is moving but may be constant in speed and direction on any part of your body. This is the notion of a ***stationary flow***. The flow is not changing with time at any particular point of space. I could try and follow a mote of dust in the air and would see the motion. The flow lines or integral curves will describe the wind and will be familiar to those that look at weather charts. Stationary flow along with a stationary metric and market field reflect a game symmetry.

The next chapter provides a framework for game symmetries in general and shows they play a decisive role; symmetries generate geometric structures with properties that can't be transformed away by choice of reference system.

Chapter 4

Game Symmetries

In the last chapter I proposed a complete theory of games determined by the motion of short range strategic sources that I approximate with a market fluid. Are there simple attributes in this theory? I believe the answer is yes. I argue that significant information about the theory is determined by its symmetry properties. In particular, I use the symmetry approach on the theory in the expanded choice-space and identify the metric g_{ab} and market field F^j_{ab}. I thereby obtain the equations that determine these fields in terms of the strategic sources. The flow of strategic-mass is one consequence of these equations. I therefore find that the symmetries of the theory play a decisive role and determine the form of dynamic game theory.

I begin with an examination of the symmetries of a sphere (the earth) and generalize to a discussion of what I call active and inactive strategy directions. I distinguish between covariance and isometry and then show that the symmetries determine the form of the dynamic theory. I conclude with a brief discussion of time isometry.

4.1 Earth's Symmetries

To make it clear that symmetries are helpful in understanding the theory, I return to the example of the geometry of the earth. We know that the geodesics are great circles. The geodesics capture an obvious attribute of the earth that it is a ball; it is round. However, I have not shown that the geodesic Eq. (3.4) will lead to great circle solutions. For more general shapes such as an oblate spheroid, I have provided even less insight on the geodesics other than the formal equations and an illustrative example in Fig. 2.1.

The insight we need is that roundness is a property of the large scale nature of the geometry and is reflected in related local symmetry properties of the metric or line element. The symmetry of the earth is reflected in the fact that the elements of the metric are independent of the longitude. This remains true even when more exact measurements[25] show that the earth is an oblate spheroid where the metric elements depend on latitude but not longitude.

I compute the curvature properties of the line element to see what insight they give. The metric and line element Eqs. (3.1) determine the connection components:

$$\Gamma^{\theta}_{\phi\phi} = -\sin\theta\cos\theta$$

$$\Gamma^{\phi}_{\theta\phi} = \Gamma^{\phi}_{\phi\theta} = \frac{\cos\theta}{\sin\theta} \cdot$$

All other connection components vanish. The connection components determine the curvature tensor:

$$R_{\theta\theta} = -1$$

$$R_{\phi\phi} = -\sin^2\theta\,.$$

$$R_{\phi\theta} = R_{\theta\phi} = 0$$

From this tensor, we are assured that the geometry described has curvature:

$$R = -\frac{2}{r^2}\,.$$

The resultant curvature scalar is inversely proportional to the square of the radius of curvature. The curvature scalar is small for a sphere of very large radius. This conforms to common experience that the earth is almost flat. The "energy momentum" tensor of the sphere is zero, $T_{ab} = 0$: There are no sources. The lack of sources however does not necessarily imply a lack of curvature.

---

[25] I ignore truly local effects such as mountains and wave peaks and valleys on oceans.

The connection components above provide two geodesic equations in terms of the longitude and latitude:

$$\frac{d^2\theta}{d\tau^2} - \sin\theta\cos\theta\left(\frac{d\phi}{d\tau}\right)^2 = 0$$

$$\frac{d}{d\tau}\left(\sin^2\theta\frac{d\phi}{d\tau}\right) = 0$$

At constant longitude $\dot{\phi} = 0$, there results a great circle path in which the latitude changes linearly with the parameter τ: The great circle goes through the North and South poles. For a sphere, the position of the North Pole could be at any point and so it is clear that other great circles are possible. This represents additional symmetry properties that I deal with in Appendix B [*Cf.* Eq. (B.7)]. Because of the choice of reference frame, both equations have connection contributions. The first equation displays a Coriolis force that accounts for the "spin" or motion of a non-zero longitudinal velocity.

In the second equation, I was able to combine the connection term in such a way as to display a ***conservation law***. It shows that the total derivative of something with respect to the path length is zero and so "that something" is constant or conserved along the path. Such equations do not occur by accident. They *always* occur when the metric is independent of a specific coordinate.

Consider Eq. (3.2): If the metric is independent of a coordinate, in this case ϕ, then there is a constant of motion:

$$\frac{d}{d\tau}\left(\frac{\partial L}{\partial \dot{\phi}}\right) = 0 .$$

This equation would be true also for an oblate spheroid. The result can be shown to depend on the metric and velocities along the path:

$$\frac{d}{d\tau}\left(g_{\phi b}\frac{dx^b}{d\tau}\right) = 0 .$$

The term inside the parentheses is determined by the velocities $dx^b/d\tau$ but is not numerically equal to it because of the metric factor. Moreover the inside term has transformation properties that differ from the velocity. I have noted earlier that velocities are vectors and as such are mathematical objects that transform in a specific way under linear transformations $\mathbf{U}$. The flow transforms as a quantity with an upper index. The quantity that is conserved transforms like a quantity with a single lower index, namely as $\mathbf{U}^{-1}$. The vocabulary of "upper" and "lower" makes clear the transformation distinction. For subsequent discussions I indicate by the term ***flow*** a velocity (upper index) and the separate term ***charge*** for its lower index relative. To each flow there is a corresponding ***conjugate*** charge: They represent the same geometric structure. They are conjugate in that they transform oppositely. Moreover their contracted product (the length of the structure) transforms as a scalar. The above conservation law is that the **charge conjugate to the flow along the longitude direction is conserved.**

This conservation law is a novel way to state a theorem in spherical geometry. On a sphere, the notion of a conserved charge associated with rotations about the polar axis is the geometric fact normally called the Law of Sines[26]:

$$\sin\theta \sin\alpha = \text{constant}.$$

It holds along any great circle. Between any two points there is a great circle so this relates the latitude and bearing α at the two points:

$$\frac{\sin\theta_1}{\sin\theta_2} = \frac{\sin\alpha_1}{\sin\alpha_2}.$$

The relationship with the conservation law follows from the identification of the bearing with the flow along the two orthogonal geodesics specified by the latitude and longitude:

$$\sqrt{g_{\theta\theta}}\frac{d\theta}{d\tau} = \cos\alpha$$
$$\sqrt{g_{\phi\phi}}\frac{d\phi}{d\tau} = \sin\alpha$$

[26] There are of course many mathematical texts on spherical trigonometry. However, it is of interest to sailors that the calculation comes under the heading of celestial navigation and therefore something useful. The calculations are contained in "sight reductions", taking sightings and reducing them to positions on the earth. See any handbook on celestial navigation. I found an interesting one, Letcher (1977).

With the appropriate line element, this applies to an oblate spheroid[27]. For a sphere, the relationship is

$$\sin\theta \frac{d\phi}{d\tau} = \sin\alpha .$$

The conserved longitude–charge is $\sin\alpha \sin\theta$. Although sailors struggle with the computational aspects of the Law of Sines using spherical trigonometry, its conceptual origins reflect a simple concept. When the metric is independent of a coordinate, the **charge conjugate to the flow along that coordinate direction is conserved along a geodesic.**

4.2 Active and Inactive Choices

The possibility of conserved charges along a geodesic suggests useful distinctions for games. I define ***inactive*** choices as those associated with a conserved charge. A choice that is not inactive is ***active***. If the choice corresponds to a strategy–choice (as opposed to a time–choice or value–choice), I call the corresponding strategy inactive or active according to the property of the strategy–choice. A discussion of these properties will help in the dynamic theory of games by providing insight into the large scale behaviors that result from local geometric structures. Of particular interest are classes of geometries in which certain coordinates or choices are always inactive. For example the ***Value–Choice Hypothesis*** can be expressed by the assertion that the value–choice ξ^j of player j is inactive. The Stationary Hypothesis is the assertion that the time–choice is inactive.

Differential geometry makes precise the notion of active and inactive in a way that is independent of coordinate system. The first step is to define a coordinate as inactive if there exists some reference frame in which the metric is independent of this coordinate. The next step is to express this in a way that is independent of coordinate system. The powerful result from differential geometry is that the existence of such a coordinate system for each inactive coordinate is equivalent to the

---

[27] The line element for a sphere is modified by allowing the radius to be a function of the latitude: $r(\theta)\sin\alpha\sin\theta$ is the conserved charge.

existence of an ***inactive vector field*** K^{μ} with certain specified properties. This vector field determines an integral curve through each point in the direction along which the metric is locally constant. I outline this approach in Appendix B. The conclusion is that the symmetry of the metric is determined by the complete set of inactive vector fields. These symmetries in turn determine charges that are conserved along a geodesic. They also determine conserved currents when external sources are present.

4.3 Covariance or Isometry

In the study of spaces where the metric is invariant under a symmetry group, it is helpful to distinguish transformations which leave the form of an equation such as the line element unchanged but change the functions that appear and transformations that leave the value of the function unchanged. Of particular importance in differential geometry is the "inner product" or line element formed from the metric $ds^2 = g_{ab}dx^a dy^b$. Any general linear transformation of the differential dx^a leaves the relationship between path length, metric and differential unchanged. The relationship is "***covariant***". The numerical value of the path length is ***invariant***, having the same value in all frames.

Not all linear transformations change the components of the metric. Those that do not are called isometries. In terms of the metric, an ***isometry*** transformation $\mathbf{D}$ is one which satisfies $\mathbf{DgD}^T = \mathbf{g}$. Such transformations are called ***orthogonal***. The set of matrices which satisfy this property for a given metric determine the orthogonal group of transformations under which the metric space is invariant. The translations $\mathbf{D} = \exp \lambda \mathbf{X}$ are determined by a vector $\mathbf{X} = \partial$, an operator on the space. The orthogonality condition $\mathbf{DgD}^T = \mathbf{g}$ for this operator implies that the metric is independent of the corresponding coordinate. Thus a coordinate is inactive if the translation operator for this coordinate is orthogonal.

The Value–Choice Hypothesis is that the translation along the value–choice is an orthogonal transformation and hence an isometry. If the metric were independent of time, then the operator formed by translations along the time direction would be an orthogonal transformation and therefore an isometry. Rotations also determine an isometry: $\mathbf{D} = \exp \theta \mathbf{J}$: $\mathbf{J}^{\mathrm{T}} = -\mathbf{J}$. The requirement that such a transformation be an isometry is that $\mathbf{DgD}^T = \mathbf{g}$. The requirement is satisfied if there is an antisymmetric tensor field $J_{ab} = -J_{ba}$. In essence, a rotation is a translation in an angular coordinate system. With these distinctions in mind I identify some candidate isometries.

4.4 Dynamic Theory of Games

The conclusion about symmetry is far reaching for game theory: The metric is the determinant of the theory. The ***isometries*** (the transformations that leave the metric unchanged) determine a local algebraic structure and hence a local group structure, that constrains the form of the metric. The large scale behavior follows from this local algebra.

As a first step in identifying the game symmetries, I deal with an assertion made in the Introduction: If I expand the geometry to include an inactive coordinate for each player, the value–choice and I assume a perfect fluid, then I recover the appropriate equations for games described by a market field and a metric. For numerical analysis I will focus on zero-sum games and make the ***Common–Value–Choice Approximation*** that those games have a common value–choice coordinate ξ^0.

I start in the extended expanded choice–space with a general split of strategies between active and inactive described in Appendix B. The line element is Eq. (B.9):

$$ds^2 = \gamma_{jk}\left(d\xi^j + A^j_a dx^a\right)\left(d\xi^k + A^k_b dx^b\right) + g_{ab}\, dx^a\, dx^b . \qquad (4.1)$$

The line element reflects the split between the inactive value–choice coordinates and the active strategy–choice coordinates; its consequences are described in more detail in Appendix C. In this section, I assume the time–choice coordinate is active. I identify the metric with g_{ab}, the flow with V^a and the market field F^j_{ab} for player j with $F^j_{ab} = \partial_a A^j_b - \partial_b A^j_a$.

I express the results in terms of a renormalized flow that is normalized to unity. The flow V^a is part of the extended mixed–choice–space, Eq. (C.9):

$$\gamma^{jk}V_jV_k + g_{ab}V^aV^b = 1. \tag{4.2}$$

With Eq. (4.1), the source Eq. (3.7) splits into three sets of equations: inactive geometry equations, active geometry equations and a mix. The "mix" Eq. (C.1) determines the market field in terms of a current:

$$\begin{gathered}\tfrac{1}{2}\frac{1}{\sqrt{|g\gamma|}}\partial_b\left(\sqrt{|g\gamma|}g^{ac}g^{bd}\gamma_{jk}F^k_{cd}\right) = \kappa\left(\mu + p\right)V_jV^a \\ F^k_{cd} = \partial_c A^k_d - \partial_d A^k_c, \quad \gamma = \det\gamma_{jk}, \quad g = \det g_{ab}\end{gathered} \tag{4.3}$$

The importance of this set of equations is that it shows that the market fields are like the Electro–Magnetic Fields in physics. They are derived from a ***messenger potential***, the ***Market Potential*** and have as their source a ***Market Current*** determined by the fluid flow ρV^a and a ***conserved charge:***

$$c_j = \frac{\mu + p}{\rho}V_j .$$

The market field equations are a bonus, since they were not considered at the outset but they are reasonable and the analog of Maxwell's equations. If the fluid has a charge in addition to its elastic properties, then fluid flow itself generates a current and modifies the field.

The active geometry Eqs. (C.4) generalizes Eq. (1.2):

$$R_{ab} - \tfrac{1}{2}g_{ab}R = \begin{pmatrix} -\kappa\left(\left(\mu+p\right)g_{ac}g_{bd}V^cV^d - pg_{ab}\right) \\ -\tfrac{1}{2}\gamma_{jk}F^j_{ac}F^k_{bd}g^{cd} + \tfrac{1}{8}g_{ab}g^{qc}g^{pd}\gamma_{jk}F^j_{pc}F^k_{dq} \\ -\tfrac{1}{2}\gamma^{jk}\gamma_{jk;ab} - \tfrac{1}{4}\gamma^{jk}{}_{;a}\gamma_{jk;b} + \tfrac{1}{2}g_{ab}g^{cd}\gamma^{jk}\gamma_{jk;cd} \\ +\tfrac{1}{2}g_{ab}g^{cd}\left(\tfrac{1}{4}\gamma^{mn}\gamma^{jk} - \tfrac{3}{4}\gamma^{jm}\gamma^{nk}\right)\gamma_{mn;c}\gamma_{jk;d} \end{pmatrix}. \tag{4.4}$$

These equations provide the active metric components with the addition of market field terms $\tfrac{1}{2}\gamma_{jk}F^j_{ac}F^k_{bd}g^{cd}$ and inactive metric component terms $\tfrac{1}{2}\gamma^{jk}\gamma_{jk;ab} + \tfrac{1}{4}\partial_a\gamma^{jk}\partial_b\gamma_{jk}$. The inactive metric components γ_{jk} act like "scalar" fields and provide the structural coupling. The active geometry is determined by the energy momentum from the elastic fluid, the energy of the market field and the energy of the scalar fields. As promised, the active geometry sources specifically include the market fields.

The inactive geometry Eqs. (C.3) determine the scalar fields:

$$\begin{aligned}&\frac{1}{\sqrt{|g\gamma|}}\gamma_{ik}\partial_a\left(\sqrt{|g\gamma|}g^{ab}\gamma^{kl}\partial_b\gamma_{lj}\right)=\\&\left(-2\kappa\left((\mu+p)V_iV_j-\frac{\mu-p}{D-2}\gamma_{ij}\right)+\tfrac{1}{2}\gamma_{ik}\gamma_{jl}F^k_{ac}F^l_{bd}g^{ab}g^{cd}\right)\end{aligned} \quad . \tag{4.5}$$

The sources for the scalar fields include the external fluid forces as well as the market field. A complete solution of the field equations will necessarily have such scalar fields.

The above equations contain the constraints on the external sources due to the conservation laws Eq. (3.8). Of all the equations, these are perhaps the ones of most interest initially. They determine the behavior of flow and charge in the model. I determine these equations in Eq. (C.8):

$$\begin{aligned}&\frac{DV^a}{\partial\tau}=V^a{}_{;b}V^b\\&\frac{DV^a}{\partial\tau}=g^{ab}V_kF^k_{bc}V^c+\left(g^{ab}-V^aV^b\right)\frac{\partial_b p}{\mu+p}-\tfrac{1}{2}g^{ab}V_jV_k\partial_b\gamma^{jk}\end{aligned} \quad . \tag{4.6}$$

These are the flows I associate with choice–space and time. The acceleration depends on the market field, on the fluid pressure and on the scalar field.

For each player there is a conserved charge determined by the player value–choice:

$$c_j=\frac{\mu+p}{\rho}V_j \, . \tag{4.7}$$

Equation (C.6) shows that c_j has a constant value along a streamline and is a conserved quantity. It appears in Eq. (4.6) like a charge. Previously, I noted that the charge $\hat{\gamma}_{j\mu}dx^\mu/d\tau$ is constant along a geodesic. The component V_j, due to the external pressure (or control), is in general not equal to this charge. I note however that if the pressure vanishes, then V_j is conserved. This is consistent with Eq. (4.7) since when the pressure vanishes, the energy and market density are proportional. Therefore calling c_j a conserved charge seems appropriate and the reasonable generalization of the naming convention when the motion is not along a geodesic.

In addition to this conserved charge, there is a consistency equation for the conservation of strategic-mass, Eq. (C.5):

$$\partial_a\left(\rho\sqrt{\gamma g}V^a\right)=0\,. \tag{4.8}$$

I now have a complete set of dynamic equations for the theory of games. They hold for any game satisfying the Value–Choice Hypothesis.

4.5 Time Isometry

In the remainder of this chapter, I consider a possible sub-class of games based on assuming time is an isometry. The simplest flows are those that are stationary. Is there reason to believe that stationary games might form an interesting sub-class? To answer this question I consider a multi-person game and suppose that for each player, there is a ***boundary position*** specified by a constant flow vector $\{\zeta^m \quad m_0\}$ and a constant market field (game matrix) f^j_{ab}. These might be the values specified at the ***defensive position*** or at a ***coalition position***. I further assume that the flow vector is in the null space of the market field: $f^j_{ab}\zeta^b=0$. I explore what isometries can be expected based on the behaviors around this position.

I start with the space and time components of the market potential[28]:

$$A^j_m=-\tfrac{1}{2}f^j_{mn}y^n+\cdots,\quad A^j_0=\frac{1}{m_0}\zeta^m f^j_{mn}y^n+\cdots. \tag{4.9}$$

[28] I use ***strategic scalars*** that are zero along the boundary point. For example, they might be related to the strategy–choices by $y=x-x^{-1}$. For large strategy–choices, the two variables are equal; for vanishingly small strategy–choices, the coordinate becomes large and negative. The coordinate is zero when the strategy–choice is unity. I imagine such a change of variable for each coordinate except time.

I have considered the behavior in the neighborhood of this boundary position. I check that the formulae provide the expected form for the market fields:

$$\begin{aligned} F^j_{mn} &= \partial_m A^j_n - \partial_n A^j_m = f^j_{mn} + \cdots \\ F^j_{m0} &= \frac{1}{m_0} \zeta^n f^j_{nm} + \cdots \end{aligned} \quad . \tag{4.10}$$

These definitions lead to the result that the flow is in the null space of the market field at the boundary position:

$$\begin{aligned} F^j_{mn}\zeta^n + m_0 F^j_{m0} &= f^j_{mn}\zeta^n + \zeta^n f^j_{nm} = 0 \\ F^j_{0m}\zeta^m &= \frac{1}{m_0} \zeta^n f^j_{nm}\zeta^m = 0 \end{aligned} \quad .$$

The proposed form of Eq. (4.9) suggests that time could be an isometry. There is an ambiguity however. I can choose a variety of coordinate frames in which to describe the game. There is a set of orthogonal transformations, isometries, which leave the metric unchanged. Any one of these will produce a stationary market field. Any such stationary market field can be described by a stationary potential. The question is which frame to choose?

I note however that there is one frame in which the payoff f^j_{mn} has the game–theory meaning. A natural choice for frame would be the one in which the payoff matrix is specified in terms of the given pure strategies. This choice is made in Eq. (1.8). I call this the ***player frame***, since it is the one in which each player α has a specific set of ***strategy–choices*** $x^{a(\alpha)}$ assigned. In the player frame, I can define a market field, market potential and metric, which are independent of time.

I conclude that it is at least plausible that there is an interesting sub-class of models in which translation in time is an isometry. I further assume that it is central [*Cf.* Appendix C] with the hidden player symmetry. I then use Appendix C to determine the resultant form of the equations. The form will be unchanged from the discussion in the previous section; I expand the inactive space to include the time choice. I note there will be an additional conserved charge c_0 associated with the time–choice. Of course the symmetry attributes associated with time will be implicit in the equations derived in the last section if the metric and sources are stationary [*Cf.* Eq. (5.6)]. Both approaches yield the same result.

I call the above statement the ***Stationary Hypothesis***: Time is an inactive choice and is central with the player value–choices. Based on this assumption, it is possible to show that there exists an inactive vector field K_μ associated with time–choice [*Cf.* Eq. (B.4)]. The length of this field $\phi^2 = \hat{\gamma}_{\mu\nu} K^\mu K^\nu$ is a scalar. It can be shown that this scalar field can be identified with a "gravitational" field [*Cf.* the ***centrally co-moving hypothesis*** systems described for an exact solution in Appendix D, Sec. D.1]. Thus there is a definition of ***gravity*** that comes out naturally in the dynamic theory of games. The stationary hypothesis is appealing because it introduces the concept of the gravitational field as a geometric structure, though I point out that the theory of games does not rely upon this assumption.

Chapter 5

Analysis

In the last chapter, I proposed a specific dynamic theory of games [Sec. 4.4] that is general, covers any number of players and extends the static theory of Von Neumann and Morgenstern (1944). I motivated the theory with examples from geometry, from symmetry arguments based on our knowledge of the geometry of earth and from examples from the extensive literature of differential geometry, fluid dynamics and physics. However, the theory must stand or fall on its own. Without recourse to the specific motivations that suggested the theory, I inquire into the behavior that results and judge whether the theory is a reasonable start towards a dynamic understanding of economic behavior. For this inquiry I take guidance from other fields on the solution and analysis domain. A reader not interested in the mathematical analysis can skip this chapter, relying on the graphical presentation of the next chapter to provide a feel for the form of the solutions.

Like many theories from the physical sciences, the predictions of the theory are not easily obtained. The mathematics or numerical machinery is not yet up to providing ready solutions to such equations. I follow an approach familiar to students of the physical sciences. I look for examples that are exactly soluble, for analyses based on the nature of the exact solutions that provide characteristic behaviors and for numerical examples that illustrate key behaviors.

I start with Appendix D where I provide a technical exposition for a completely soluble model with a single active strategy. Despite its limitations this model provides critical insight into the nature of the equations. For a first reading, this appendix can be skipped; in so doing the reader will have to accept the results quoted in this chapter. Next, I use the covariance of the geometry to identify reference systems in which useful approximations can be made that are consistent with the exactly soluble model and consistent with what insight we have from the static theory. I provide a set of models based on the line element Eq. (5.4) that admits analysis and numerical solution for the equations of motion. I show with a change of variables the simplified set of flow equations, Eq. (5.5) that result. I analyze their ***basic behaviors*** in Sec. 5.4. These basic behaviors nicely characterize many of the graphical behaviors presented in the next chapter.

5.1 Reference Frames

I have assumed the Value–Choice Hypothesis and the Stationary Hypothesis. In search of a simplified line element, I inquire into other isometries. The easiest isometries to deal with are those that commute with each other [Appendix C]. An extreme example of this would be a model in which there is a single active strategy and all inactive choices are central. I call this the Single Strategy Model and solve this model in Appendix D. In particular, I derive the temperature within this model. Numerical examples provide additional and alternative insight and are provided in Appendix E. These models are based on a complete solution to all the field equations.

I have not identified other examples of a complete solution. In lieu of such other solutions, I focus on the equations of motion Eq. (4.6) where I ignore the equations for the metric and the conservation of energy and strategic-mass but take an assumed form for the line element and endow it with symmetry and/or isometry properties. A proposal for considering the conservation laws is given in Appendix F. In these models I assume that the structural couplings γ_{jk} for the inactive choices are constants. If each player sees the same market field, there will be no loss of generality taking the ***Common–Value–Choice Approximation***. I note that an analysis can also be done in which each player sees a different market field. This covers games such as the prisoner's dilemma (see Sec. 7.6).

Starting from the static theory, I investigate isometries that might hold, even if they are not generally true in all models[29]. At a boundary position, the market potential has a form that insures that the market field has a recognizable form in which there are no ***internal factions***:

$$F_{m(\alpha)n(\alpha)} = 0 .$$

There is no guarantee that the field equations will maintain this principle. I call this the ***player form*** for the market field. I believe a necessary and sufficient condition for this is that the market field be expressed in terms of scalars as follows:

$$\begin{aligned} A_{m(\alpha)} &= \partial_{m(\alpha)}\psi_\alpha \\ A_0 &= \psi_0 \end{aligned} . \tag{5.1}$$

Though it is not an isometry, the player form insures that there are no internal factions. However there may be isometries of the metric and they will be reflected in the behavior of the N scalars ψ_α ψ_0. I use this form for the graphical presentation in Chapter 6. I emphasize that the player form though useful is a provisional step towards gaining an understanding of the equations of motion. In the end, the theory provides a full set of equations which dictate the form of the metric based on the sources and so may well be inconsistent with this assumption.

In the vicinity of a boundary position ζ^a, the ***player payoff*** between players is proportional to [*q.v.* Footnote 28]

$$\Delta^{(\alpha,\beta)} = y^{a(\alpha)} F_{a(\alpha)b(\beta)} y^{b(\beta)} . \tag{5.2}$$

In numerical examples, I use this form to compute how much a player receives from each other player. This form is crucial to establishing a link between the theory and observation. I evaluate the invariant payoffs using strategy–choices normalized by the ***player strategic length***

$$\sum_m x^{m(\alpha)} .$$

---

[29] Our familiar example of the earth is suggestive. The earth is not exactly a sphere but that symmetry provides an interesting starting point.

If all the strategy–choices are sufficiently large, this length is approximated by

$$\sum_m y^{m(\alpha)} .$$

The concept of a player is embodied in the ***player frame***. This is the frame in which the payoffs have their expected game theory values. The boundary conditions and player form specify the market potential components of the metric. Typically there will be no internal factions. However, I still have choices about how to choose the player variables. Suppose I have made one such choice as described above. I can then put the game in a standardized form I call the ***equilibrium frame***:

$$y^{m(\alpha)} = \mathbf{r}_\alpha \cdot \zeta^{m(\alpha)} + \mathbf{z}_\alpha \cdot \mathbf{q}^{m(\alpha)} . \tag{5.3}$$

I call this a ***strategic decomposition*** of the strategic scalars: For each player I introduce the ***player strategic length***

$$r_\alpha = \sum_m y^{m(\alpha)} .$$

There will be sufficient deviation vectors $\mathbf{q}^{m(\alpha)}$ which are differences of two-player strategies to form a basis with the equilibrium strategies, which I define to be the ***player deviations*** $\mathbf{z}_\alpha$. Based on the definitions I have

$$\sum_m \mathbf{q}^{m(\alpha)} = 0 .$$

This technical device provides a coordinate system in which to discuss the strategies. Most of us are familiar with street maps: Streets and addresses provide a ***coordinate system*** which allows travelers to find their destination. One of the characteristics of such maps is that they are not unique. Local residents may find it convenient to specify only a street, if for example such a street is only a block long and all local residents are familiar with that fact. That coordinate system is suggestive to those local residents. Sailors on the other hand may prefer specification of the latitude and longitude on uncharted waters, whereas the modern day sports fisherman may have a GPS that shows a map of his or her position providing quite a different specification in terms of familiar land or water marks.

5.2 Two-Person Zero-Sum Fair Game

The discussion on street maps holds for the technical device above, called the strategic decomposition. It has some advantages in demonstrating the possible behavior of the solutions against suitable landmarks called initial conditions. I illustrate the strategic decomposition and the possible advantages with an example of a two-person fair game, where each player has two strategies. There are four strategies $y^{m(\alpha)}$ corresponding to the two players. This is one possible coordinate system. The strategic decomposition provides an equivalent coordinate system:

$$y^{m(\alpha)} = r_\alpha \zeta^{m(\alpha)} + z_\alpha q^{m(\alpha)}.$$

The explicit definitions are:

$$\begin{aligned} y^1 &= r_2 \xi^1_{def} + z_2 \\ y^2 &= r_2 \xi^2_{def} - z_2 \\ y^3 &= r_1 \xi^3_{def} + z_1 \\ y^4 &= r_1 \xi^4_{def} - z_1 \end{aligned}.$$

In the new system, it is enough to know the two strategic lengths and the two-player deviations in order to determine the strategy scalars. The labels on the right correspond to player 1 and 2 and the choice on the left is one I have used for numerical calculations. The ***player frame*** variables are $\{y^m \quad t\}$. The ***equilibrium frame*** variables are $\{r_1 \quad r_2 \quad z_1 \quad z_2 \quad t\}$ corresponding to the player strategic lengths (along the equilibrium direction) and player strategic deviations ("orthogonal" to the equilibrium directions). The transformation to the new variables is:

$$S \equiv \begin{pmatrix} \partial y^a / \partial \bar{y}^b & r_1 & r_2 & z_1 & z_2 & t \\ \hline y^1 & 0 & \zeta^1 & 0 & 1 & 0 \\ y^2 & 0 & \zeta^2 & 0 & -1 & 0 \\ y^3 & \zeta^3 & 0 & 1 & 0 & 0 \\ y^4 & \zeta^4 & 0 & -1 & 0 & 0 \\ t & 0 & 0 & 0 & 0 & 1 \end{pmatrix}.$$

I note that the transformation is one that mixes player 1 with itself and player 2 with itself and allows no mixing between players. I can't yet argue whether the transformation is orthogonal relative to the metric, since I have not defined the metric; I can say at least that it is not orthogonal relative to the diagonal Minkowski metric. The transformation has determinant −1 is non-singular and has the inverse:

$$S^{-1} = \begin{pmatrix} 0 & 0 & 1 & 1 & 0 \\ 1 & 1 & 0 & 0 & 0 \\ 0 & 0 & \varsigma^4 & -\varsigma^3 & 0 \\ \varsigma^2 & -\varsigma^1 & 0 & 0 & 0 \\ 0 & 0 & 0 & 0 & 1 \end{pmatrix}.$$

By definition, the transformation determines:

$$dy^a = S^a{}_b d\overline{y}^b .$$

From this general form $d\mathbf{y} = \mathbf{S}d\overline{\mathbf{y}}$ I determine that covariant vectors transform as $\mathbf{A}^T = \overline{\mathbf{A}}^T \mathbf{S}^{-1}$. The metric in the transformed frame is $\overline{g} = \mathbf{S}^T g \mathbf{S}$. Although there is no guarantee that the transformation is an isometry, the determinant of the metric is unchanged:

$$\det \overline{g} = \det S^T \det g \det S = \det S \det g \det S = \det g .$$

A unit volume element does not change size under this transformation. The equilibrium frame is just as good a frame as the player frame.

The advantage of the equilibrium frame is that the payoff matrix $\overline{\mathbf{F}} = \mathbf{S}^T \mathbf{F} \mathbf{S}$ is put into the ***equilibrium form***:

$$\overline{F} = \begin{pmatrix} 0 & \mathrm{v} & 0 & 0 & -a \\ -\mathrm{v} & 0 & 0 & 0 & a \\ 0 & 0 & 0 & -\omega & 0 \\ 0 & 0 & \omega & 0 & 0 \\ a & -a & 0 & 0 & 0 \end{pmatrix} = \mathrm{v}\overline{\tau}_v + \omega\overline{\tau}_f .$$

The defensive strategy is the vector $\{1 \quad 1 \quad 0 \quad 0 \quad m_0\}$ and forms a null vector of the market field, so $a = \mathrm{v} m_0^{-1}$, where the game value is v. The payoff matrix depends on one parameter ω, which characterizes the payoff between the deviations. Thus the payoff matrix is the sum of two components: The ***value component*** proportional to the game value; and the fair game ***deviation component***, which is a rotation orthogonal to either player strategy.

I assert two-person zero-sum games in general can always achieve the above characteristic breakdown: A value component proportional to the game value consisting of a two dimensional block characterized by the game value and no payoffs between the deviations and the strategic lengths and the deviation component consisting of a rotation orthogonal to either player strategy.

Returning to the simplified example, I note first that the value–component has a simple form in the player frame:

$$\tau_v = \begin{pmatrix} 0 & 0 & -1 & -1 & m_0^{-1} \\ 0 & 0 & -1 & -1 & m_0^{-1} \\ 1 & 1 & 0 & 0 & -m_0^{-1} \\ 1 & 1 & 0 & 0 & -m_0^{-1} \\ -m_0^{-1} & -m_0^{-1} & m_0^{-1} & m_0^{-1} & 0 \end{pmatrix}.$$

This has the direct interpretation that I can add a constant to the payoff matrix without changing the strategic properties of the game.

I note next that I can reduce the value component to a rotation by a suitable orthogonal transformation. I define orthogonality against the metric in the equilibrium frame:

$$\bar{g}_{ab} = \begin{pmatrix} -1 & 0 & 0 & 0 & 0 \\ 0 & -1 & 0 & 0 & 0 \\ 0 & 0 & -1 & 0 & 0 \\ 0 & 0 & 0 & -1 & 0 \\ 0 & 0 & 0 & 0 & e^{2\nu_0} \end{pmatrix}.$$

I wish to transform the following matrix to a rotation:

$$\bar{\tau}_v = \begin{pmatrix} 0 & 1 & 0 & 0 & -m_0^{-1} \\ -1 & 0 & 0 & 0 & m_0^{-1} \\ 0 & 0 & 0 & 0 & 0 \\ 0 & 0 & 0 & 0 & 0 \\ m_0^{-1} & -m_0^{-1} & 0 & 0 & 0 \end{pmatrix}.$$

I consider two successive orthogonal transformations, first a rotation:

$$R = \begin{pmatrix} 1/\sqrt{2} & -1/\sqrt{2} & 0 & 0 & 0 \\ 1/\sqrt{2} & 1/\sqrt{2} & 0 & 0 & 0 \\ 0 & 0 & 1 & 0 & 0 \\ 0 & 0 & 0 & 1 & 0 \\ 0 & 0 & 0 & 0 & 1 \end{pmatrix}.$$

And second a "boost":

$$B = \begin{pmatrix} \cosh\beta_0 & 0 & 0 & 0 & e^{v_0}\sinh\beta_0 \\ 0 & 1 & 0 & 0 & 0 \\ 0 & 0 & 1 & 0 & 0 \\ 0 & 0 & 0 & 1 & 0 \\ e^{-v_0}\sinh\beta_0 & 0 & 0 & 0 & \cosh\beta_0 \end{pmatrix}.$$

The first transforms the equilibrium vector $\{1 \;\; 1 \;\; 0 \;\; 0 \;\; m_0\}$ to the vector $\{\sqrt{2} \;\; 0 \;\; 0 \;\; 0 \;\; m_0\}$ that is along a single axis; the second transforms the resultant equilibrium vector to one having only a time component: $\{0 \;\; 0 \;\; 0 \;\; 0 \;\; m_0/\cosh\beta_0\}$. The second transformation is a boost to the rest frame defined by the ***game mass*** m_0:

$$\bar{\bar{\tau}}_v = \begin{pmatrix} 0 & 1/\cosh\beta_0 & 0 & 0 & 0 \\ -1/\cosh\beta_0 & 0 & 0 & 0 & 0 \\ 0 & 0 & 0 & 0 & 0 \\ 0 & 0 & 0 & 0 & 0 \\ 0 & 0 & 0 & 0 & 0 \end{pmatrix}.$$

This is the transformed value component. The transformation leaves the fair-component unchanged. The transformation is determined by the ***equilibrium market boost*** β_0 :

$$\tanh\beta_0 = \frac{\sqrt{2}}{m_0} e^{-\nu_0} .$$

The time component of the metric g_{00} is known at equilibrium and sets the parameter ν_0: $g_{00} = e^{2\nu_0}$.

I suggest that the frame in which the equilibrium vector is at rest be the ***central frame***. It is not essential for the theory but provides an interesting class of models for numerical study. In these models, I have the possibility of seeing not only time as an isometry but rotations corresponding to the value–component and the fair–component.

5.3 Central Frame Models

These considerations lead me to consider the following class of ***central frame models***. The line element is expressed in the central frame and I assume a form:

$$ds^2 = g_{00}dt^2 + \gamma_{\xi^0\xi^0}\left(d\xi^0 + \bar{\bar{A}}_m d\bar{\bar{y}}^m + \bar{\bar{A}}_0 d\bar{\bar{t}}\right)^2 + g_{mn}\, d\bar{\bar{y}}^m\, d\bar{\bar{y}}^n . \quad (5.4)$$

In this model, I take the metric elements $\gamma_{\xi^0\xi^0} = -\varepsilon^2$ and g_{mn} to be constants. I take the space–metric components to be diagonal and equal to -1: $g_{mn} = -\delta_{mn}$. The time component of the metric g_{00} is independent of time in the central frame but otherwise an arbitrary function of the strategic scalars.

The vector potentials A_m A_0 are independent of time in the central frame and are set to reproduce a constant payoff matrix in the player frame with no internal factions. A typical form for the vector potential that I use is $A_a = -\tfrac{1}{2} f_{ab} y^b$ in the player frame, given a constant payoff matrix f_{ab}. This insures there are no internal factions. Since the various transformations above are constants, this transforms the vector potential into a stationary form $\bar{\bar{A}}_a = -\tfrac{1}{2} \bar{\bar{f}}_{ab} \bar{\bar{y}}^b$: The time component of this field vanishes. An addition to the time component of the vector potential that deviates from this I call the ***market bias***. I require the market bias to vanish at the equilibrium position.

The central frame line element is consistent with the Value–Choice Hypothesis and the Stationary Hypothesis. If the metric g_{00} and/or the market bias deviation of the time component of the vector potential A_0 have non-trivial dependence on the coordinates, they may generate internal factions.

I have a suggested simplified line element Eq. (5.4) and some possible isometries for the arbitrary functions that appear. I wish to use this line element to analyze the equations of motion, Eq. (4.6). In performing the analysis, it will be useful to make the additional ***homogeneous*** assumption that the energy density is a function of pressure. The only justification I provide at this point is that of analogy: In similar arguments from the field of fluid mechanics, such an assumption has been found useful as an initial approximation, one that does not cloud the insight obtained for the full equations. The assumption will make it possible to compare results to fluid mechanics to provide some confidence to the intuition we are trying to gain.

For central frame models, the equations of motions[30] will be shown to be Eq. (5.5); I will illustrate some of the expected symmetry properties that result. For the conserved charge associated with the common player value–choice, I make a first change of variables to a ***renormalized pressure*** φ:

$$V_{\xi^0} = c_{\xi^0} \frac{\rho}{\mu + p} \equiv c_{\xi^0} e^{-\varphi} .$$

The equation of motion for the charge shows that c_{ξ^0} is constant along the path. This renormalized pressure is determined by the pressure, since the energy density $\mu(p)$ is assumed to be a function of pressure:

$$d\varphi = \frac{dp}{\mu + p} .$$

[30] Readers not interested in the mathematical details of this exercise in changing variables can skip directly to Eq. (5.5).

The strategic-mass is determined in terms of this renormalized pressure, along with the ***market overhead index*** $\alpha \equiv d\mu/dp$:

$$\frac{d\rho}{\rho}=\frac{d\mu}{\mu+p}=\alpha d\varphi .$$

The conservation of strategic-mass can be expressed in terms of the renormalized control, the market overhead index and the ***volume compressibility*** $V^a_{;a}$:

$$\alpha\frac{d\varphi}{ds}+V^a_{;a}=0 .$$

In the numerical examples in the next chapter, I take the market overhead index α to be constant.

The equations simplify when expressed in terms of a flow normalized to unity. I start with Eq. (4.2) using the constant value $\gamma^{\xi^0\xi^0}=-\varepsilon^{-2}$ of the scalar field assumed in the player form:

$$-\varepsilon^{-2}c_{\xi^0}c_{\xi^0}e^{-2\varphi}+g_{ab}V^aV^b=1 .$$

I introduce a unit flow (with $g_{ab}W^aW^b=1$) proportional to the original flow: $V^a\equiv nW^a$. The normalization condition determines the normalization in terms of the pressure:

$$n^2=1+c_{\xi^0}{}^2\varepsilon^{-2}e^{-2\varphi} .$$

The general form of the equations of motion is simplified further in terms of a second change of variables to a new ***renormalized pressure*** ϕ defined in terms of the previous one:

$$d\phi\equiv d\varphi+\frac{1}{n}dn .$$

The resultant flow equations have the form of a transverse pressure gradient, expressed as the product of the gradient and the transverse projection operator $h_{ab}\equiv g_{ab}-W_aW_b$ and the market field contribution[31]:

$$\dot{W}_a=h^b_a\partial_b\phi+\frac{c_{\xi^0}}{n_0}e^{-\phi}F_{ab}W^b . \tag{5.5}$$

[31] Because of Eq. (5.4) the contribution from the common player value-choice vanishes.

This will be the working equation for the numerical examples. An immediate consequence of this equation is that the normalization $g_{ab}W^aW^b$ is constant along the path. Equation (5.5) consists of two terms that read like classical physics text expressions: The first is the transverse gradient of the ***renormalized pressure*** ϕ; the second term is the contribution from the market field. It has the appearance of a charge interacting with an electro–magnetic field. One change from classical physics is that the charge density has a dependence on the renormalized pressure. At large pressures, the charge density is small. The charge density is maximal when the pressure is small.

There may be other symmetries of Eq. (5.4); if present, they produce additional conserved charges. The renormalized flow equations contain the consequences of these symmetries. Thus for numerical calculations, I don't need to explicitly identify these symmetries.

From an analysis perspective however, it is interesting to see how such symmetries manifest. As an example, I consider the time isometry and omitting the algebra, give the acceleration of the time component of the renormalized flow:

$$\frac{d}{d\tau}\left(e^{\ln\cosh\beta+\nu+\phi}+\frac{c_{\xi^0}}{n_0}A_0\right)=0. \tag{5.6}$$

It is a generalization of an equation in mechanics by Euler for a charged fluid, which can be seen in the limit of small ***market boost*** β, ***gravitational potential*** $\nu=\tfrac{1}{2}\ln g_{00}$ and renormalized pressure ϕ:

$$\frac{d}{d\tau}\left(\tfrac{1}{2}\beta^2+\nu+\phi+\frac{c_{\xi^0}}{n_0}A_0\right)=0.$$

In fluid mechanics, this expresses the conservation of energy with contributions from kinetic energy (small market boost is approximately the velocity), gravitational potential, pressure and market bias forces[32]. For small pressure, the renormalized pressure is approximately $p/n_0^{\,2}\rho_0$. The fluid mechanics form is obtained by multiplying the result inside the

[32] The interested reader might find the discussion of the Bernoulli Theorem in Feynman (1963), volume 2, of interest. Excluding the market potential, we have a similar effect in higher dimensions.

brackets by the "rest" strategic-mass $n_0^{\ 2}\rho_0$, where the "electric" charge is $n_0\rho_0 c_{\xi^0}$.

Some of the readers may in fact be familiar with the more practical aspects of this equation. One manifestation of this equation is the Bernoulli equation for fluids. The flow past an airfoil or sail operates on this principle; the faster the flow on one side of a foil, the lower the pressure. The resultant pressure difference causes the foil to lift. This provides an understanding of why airplanes fly and why sailboats are able to sail close to the wind.

Omitting the algebraic detail, the acceleration of the space components is:

$$g_{mn}\frac{dW^n}{d\tau} = \begin{pmatrix} \left(\cosh^2\beta\right)\left(\partial_m \nu\right) + \left(\delta_m^n - W_m W^n\right)\partial_n \phi \\ +\dfrac{c_{\xi^0}}{n_0} e^{-\phi} F_{mn} W^n + \dfrac{c_{\xi^0}}{n_0} e^{\ln\cosh\beta - \phi - \nu}\partial_m A_0 \end{pmatrix}. \tag{5.7}$$

It can be shown that Eq. (5.6) is a consequence of these equations. The interpretation of the terms in Eq. (5.7) is instructive. We are in the central frame. The left hand side is the acceleration of strategic-mass in the central frame for a play along a curve, a sequence of plays. The first term on the right is a "gravity" contribution that depends weakly on market boost $\cosh^2\beta \approx 1+\beta^2$ and the gravitational potential ν. It shows that a potential–well causes an attraction, justifying the metaphor of gravity. The second term is the transverse gradient of the renormalized pressure. It reflects a concept familiar in meteorology and sailing that the wind moves from high pressure to low pressure. The remaining two terms are the market contributions. The first of these is the influence of the payoffs to the flow. The last of these reflects the market bias.

5.4 Basic Behavior

The dynamic theory of games is both a geometric and mechanical theory characterized by a metric and an elastic market medium, Eq. (1.2). The theory determines the behavior of the sources Eq. (3.8). The elastic medium obeys the conservation of strategic-mass and obeys a force equation that relates acceleration of the flow of the substance to the market pressure gradient. The analogy to weather that I have noted is striking and insightful. The flow determines and helps describe how we might view markets and economic behaviors.

In Sec. 5.3 I proposed Eq. (5.4) as a suitable class of solutions for initial analysis. It presupposes the Value–Choice Hypothesis, the Stationary Hypothesis, the Common–Value–Choice Approximation and the assumption that the form of the metric components are set in the player frame. There are three functions that can be varied: pressure (or control), market bias and gravitational potential. In a more complete solution (as for example in the exact solution in Appendix D), these quantities are computed from the full set of field equations. Equation (5.7) complements the line element by showing the influence of the elastic media on the flow.

In this section, I focus on the effects of the game matrix contribution. I find that a general characteristic of the equations is a periodic or near periodic behavior I call ***oscillatory***, which is a consequence of the game matrix. To see this, I ignore the pressure and gravity effects and recall that the market field is assumed to be constant.

In this case, using appropriate units for the proportionality constants, the equation of motion Eq. (5.5) is a ***pure game flow***:

$$g_{ab}\frac{dW^b}{d\tau} = F_{ab}W^b . \tag{5.8}$$

The form of the solutions can be obtained in their entirety. First, I note that given two solutions, their linear combination will again be a solution. In general, any linear combination or superposition of solutions is again a solution. Second, I note an important class of solutions associated with ***null vectors*** ζ^a of the constant game matrix, defined as those necessarily constant vectors satisfying

$$F_{ab}\zeta^b = 0 . \tag{5.9}$$

These null vectors are solutions to the pure game flow: They are the ***null solutions.*** Strategy–choices and time associated with such a null vector will grow linearly with the path length parameter τ. In two-person games, such null vectors are identified with the stationary game equilibrium. The interpretation is less clear for games with three or more players, since the assumption of a constant game matrix may not be appropriate.

The remaining solutions to the equation can now be identified. The game matrix is an anti-symmetric matrix and generates rotations in the space–time of the theory. The charge conjugate to this rotation (called "angular momentum" in physics texts) is conserved. In addition, any anti-symmetric matrix that commutes with the game matrix generates a rotation whose conjugate charge is also conserved. These results can be shown directly.

A way to obtain these results is to note that in a metric space, given any anti-symmetric matrix, in addition to there being a set of null vectors, there will always be a complete set of pairs of ***rotational eigenvectors*** $\left\{u^a_\omega \quad v^a_\omega\right\}$ labeled by an ***eigenvalue*** ω with the following properties:

$$F_{ab}u^b_\omega = \omega g_{ab}v^b_\omega, \quad F_{ab}v^b_\omega = -\omega g_{ab}u^b_\omega . \tag{5.10}$$

This states the fact that in an appropriate basis, any anti-symmetric matrix is a finite sequence of zero 1×1 matrices corresponding to the null vectors and 2×2 anti-symmetric matrices, whose off-diagonal term is specified by its eigenvalue ω. The number of such eigenvalues is necessarily finite. This is supported by the following normalizations for the eigenvectors:

$$\begin{aligned} &g_{ab}u^a_\omega u^b_\omega = g_{ab}v^a_\omega v^b_\omega = -1, \quad g_{ab}u^a_\omega v^b_\omega = g_{ab}u^a_\omega \zeta^b = g_{ab}v^a_\omega \zeta^b = 0 \\ &\omega \neq \omega' \Rightarrow g_{ab}u^a_\omega u^b_{\omega'} = g_{ab}u^a_\omega v^b_{\omega'} = g_{ab}v^a_\omega v^b_{\omega'} = 0 \end{aligned} .$$

A solution to the pure game flow can be constructed from these eigenvectors:

$$A^a_\omega = u^a_\omega \cos\omega\tau + v^a_\omega \sin\omega\tau .$$

The verification can be made by computing both sides of the equation:

$$\begin{aligned} &g_{ab}\frac{dA^b_\omega}{d\tau} = -\omega g_{ab}u^b_\omega \sin\omega\tau + \omega g_{ab}v^b_\omega \cos\omega\tau \\ &F_{ab}A^b_\omega = \omega g_{ab}v^b_\omega \cos\omega\tau - \omega g_{ab}u^b_\omega \sin\omega\tau \\ &\therefore g_{ab}\frac{dA^b_\omega}{d\tau} = F_{ab}A^b_\omega \end{aligned} .$$

Similarly there is an orthogonal solution, whose verification is similar:

$$B^a_\omega = -u^a_\omega \sin\omega\tau + v^a_\omega \cos\omega\tau .$$

I summarize the rotational solutions:

$$\begin{aligned} A^a_\omega &= u^a_\omega \cos\omega\tau + v^a_\omega \sin\omega\tau \\ B^a_\omega &= -u^a_\omega \sin\omega\tau + v^a_\omega \cos\omega\tau \end{aligned} . \tag{5.11}$$

The general solution will be an arbitrary linear combination of the null and rotational eigenvalue solutions.

The flow determines the variation of a coordinate through the ***integral curve***:

$$\frac{dy^a}{d\tau} = W^a .$$

It can be verified that the general solution for the coordinate is also determined by the null and rotational eigenvector solutions[33]:

$$y^a = a^a_0 + \tau b_0 \zeta^a + \sum_\omega \left(a_\omega A^a_\omega + b_\omega B^b_\omega \right), \tag{5.12}$$

I differentiate this with respect to the path length to obtain the flow and so verify that this is the general solution:

$$W^a = b_0 \zeta^a + \sum_\omega \omega \left(a_\omega B^a_\omega - b_\omega A^b_\omega \right). \tag{5.13}$$

The basic behavior solution is expected to hold whenever the gravitational and pressure contributions to the equations of motion can be ignored. Thus in general I expect the coordinates to have a linearly rising component along with oscillating components that consist of a superposition of terms with different frequencies. Gravity and pressure will modify this basic behavior.

The basic behavior holds for games with any number of players and I expect remnants of this behavior to appear even in the presence of gravity and pressure that destroy the angular symmetry. However, for two-person games, under certain circumstances the symmetry might be

[33] For simplicity I show only a single null solution: There may in fact be several and so there will be a sum of such contributions.

exactly preserved and a generalization of the above result obtained valid in the presence of both a gravitational field and a non-zero pressure.

To see this I consider the following form for the market potential in the central frame which determines a constant payoff matrix and is consistent with Eq. (5.4):

$$\overline{\overline{A}}_a = -\tfrac{1}{2}\overline{\overline{f}}_{ab}\overline{\overline{y}}^b .$$

They determine the payoff matrix components $F_{ab} = f_{ab}$. The mixed time components are chosen such that the equilibrium vector $\{\zeta^m \quad m_0\}$ is in the null space of the payoff matrix: $F_{ab}\zeta^b = 0$.

Earlier [*Cf.* the discussion in Sec. 5.2] I have demonstrated that the strategic decomposition variables split the payoff matrix into two components: The value component and the fair–component. In the central frame, the value component is in a block diagonal form for the subspace of the strategic lengths:

$$\overline{\overline{\tau}}(\omega_{\text{v}}) = \begin{pmatrix} 0 & -1 \\ 1 & 0 \end{pmatrix}.$$

The fair–components can be similarly put into a block diagonal form (with possibly some number of pairs of null 1×1 blocks). I assume that the central frame is in this block diagonal form. Each block is given in terms of a 2×2 matrix for the appropriate subset of strategies:

$$\overline{\overline{\tau}}(\omega) = \begin{pmatrix} 0 & -1 \\ 1 & 0 \end{pmatrix}.$$

Equation (5.4) can be expressed in this basis using the form for the market potential:

$$ds^2 = -\varepsilon^2\left(d\xi^0 - \tfrac{1}{2}\overline{\overline{f}}_{mn}\overline{\overline{y}}^n d\overline{\overline{y}}^m\right)^2 + g_{mn}\, d\overline{\overline{y}}^m\, d\overline{\overline{y}}^n .$$

The market field is represented by the sum over the 2×2 matrices in the central frame:

$$\overline{\overline{f}}_{mn} = \sum_{\omega} \omega \overline{\overline{\tau}}(\omega)_{mn} .$$

The new coordinates occur in pairs associated with the 2×2 matrix for each eigenfrequency:

$$g_{ab}d\bar{\bar{y}}^a d\bar{\bar{y}}^b = g_{00}d\bar{\bar{t}}^2 - \sum_{\omega}\left(d\bar{\bar{u}}_\omega^{\,2} + d\bar{\bar{v}}_\omega^{\,2}\right).$$

From this I can simplify the line element contribution from the market potential:

$$\bar{\bar{f}}_{mn}\bar{\bar{y}}^n d\bar{\bar{y}}^m = \sum_{\omega}\omega\left(d\bar{\bar{y}}^m\tau(\omega)_{mn}\bar{\bar{y}}^n\right) = \sum_{\omega}\omega\left(\bar{\bar{u}}_\omega d\bar{\bar{v}}_\omega - \bar{\bar{v}}_\omega d\bar{\bar{u}}_\omega\right).$$

The line element has the form:

$$ds^2 = \left\{\begin{aligned} &-\varepsilon^2\left(d\xi^0 - \tfrac{1}{2}\sum_{\omega}\omega\left(\bar{\bar{u}}_\omega d\bar{\bar{v}}_\omega - \bar{\bar{v}}_\omega d\bar{\bar{u}}_\omega\right)\right)^2 \\ &-\sum_{\omega}\left(d\bar{\bar{u}}_\omega^{\,2} + d\bar{\bar{v}}_\omega^{\,2}\right) - d\bar{\bar{y}}_\varsigma^{\,2} \end{aligned}\right\}.$$

To explicitly identify the inactive strategy, I transform each pair that is not null vectors to polar coordinates, which provide an angular and radial strategy for each 2×2 matrix:

$$ds^2 = \left\{\begin{aligned} &g_{00}d\bar{\bar{\bar{t}}}^2 - \varepsilon^2\left(d\xi^0 - \tfrac{1}{2}\sum_{\omega}\omega\bar{\bar{\bar{r}}}_\omega^{\,2} d\bar{\bar{\bar{\theta}}}_\omega\right)^2 \\ &-\sum_{\omega}\left(d\bar{\bar{\bar{r}}}_\omega^{\,2} + \bar{\bar{\bar{r}}}_\omega^{\,2} d\bar{\bar{\bar{\theta}}}_\omega^{\,2}\right) - d\bar{\bar{\bar{y}}}_\varsigma^{\,2} \end{aligned}\right\}. \tag{5.14}$$

For the value component, I name the respective strategies the ***angular value strategy*** and the ***radial value strategy***. For the deviation component, I name the respective strategies the ***angular deviation strategy*** and the ***radial deviation strategy***.

If the time component of the metric g_{00} is a function only of the radial strategies, then the angular strategies will be inactive strategies. I thus identify an interesting class of two-person zero-sum dynamic games where the only active choices are the radial strategies. Moreover, the inactive choices, including time and the common player value–choice, are central. Therefore, their conservation properties are covered by the results in the Appendix C. In particular, I obtain the conserved constant

c_{θ_ϖ} along a streamline assuming the specific form above for the line element:

$$-\overline{\overline{\overline{r}}}_\varpi^{\,2} \frac{d\overline{\overline{\overline{\theta}}}_\varpi}{ds} = \overline{\overline{\overline{V}}}_{\theta_\varpi} - \overline{\overline{\overline{A}}}_{\theta_\varpi} \overline{\overline{\overline{V}}}_{\xi^0} = \frac{\rho}{\mu + p}\left(c_{\theta_\varpi} - \overline{\overline{\overline{A}}}_{\theta_\varpi} c_{\xi} \right).$$

I convert back to the central frame and then to the player frame to get the result for conserved angular momentum:

$$c_{\theta_\varpi} = -\left(\frac{\mu + p}{\rho} \frac{dy^a}{ds} + c_{\xi^0} A^a \right) \tau(\omega)_{ab}\, y^b . \qquad (5.15)$$

It is not gauge invariant but is invariant under constant coordinate transformations and is valid in the presence of the scalar fields, pressure and a market bias. In particular, the market field determines a conserved angular momentum:

$$c_{\theta_\varpi} = -\left(\frac{\mu + p}{\rho} \frac{dy^a}{ds} + c_{\xi^0} A^a \right) f_{ab}\, y^b .$$

This will be a linear combination constructed from Eq. (5.15).

I conclude that when the unknown metric and market components are functions only of the radial directions, then the basic behavior Eq. (5.12) generalizes and holds in the presence of pressure and gravity. In these cases, there will be appropriate conserved charges corresponding to associated angular isometries. However, in general linear combinations of solutions will no longer generate new solutions.

I also note that for fair games, I generate two translational symmetries in the central frame: Both player strategic lengths can be taken as inactive. For the special case of a fair game where each player has only two strategies, I can construct a game with a single active strategy: The two strategic lengths, time and the angular deviation strategy are inactive.

Chapter 6

Graphical Presentation

In chapter 4, I proposed a specific dynamic theory of games that is general, covers any number of players, covers zero-sum and non-zero sum games and extends the static theory of Von Neumann and Morgenstern (1944). I motivated the theory by examples from geometry, from symmetry arguments based on our knowledge of the geometry of earth and from examples from the extensive literature of differential geometry, fluid dynamics and physics. However, the theory stands or falls on its own. Without recourse to the specific motivations that suggested the theory, I inquire into the behavior that results and judge whether the theory is a reasonable start towards a dynamic understanding of economic behavior. In this chapter I provide a graphical presentation of the behavior expected for both two-person and three-person games[34]. The mathematical analysis was presented in the previous chapter.

I recall that the goal of this monograph is to help the practice of strategy be a science, not an art. What distinguishes science from art in part is based on measurement: Its ability to make quantitative statements. For example, science goes further than saying space flight is possible; predictions can be made for the trajectory of a proposed spaceship; a

[34] I make light of the work necessary to present the result of the equations in graphical form. However the effort has not been small. Some of the work however has been made easier due to computer programs on the market. I have used *Mathematica*® by Wolfram (1992). Without such support, the calculations would have been tedious and even more error prone. I also note the existence of a Systems Dynamics program iThink® by High Performance Systems (1997). I have cross checked some of the calculations with this, as well as used it for systems dynamics calculations to generate insight for some of the results. Though hidden, it goes without saying that these numerical packages have been essential to obtaining insight into the equations.

trajectory determines the position of the ship as a function of time. It is not as picturesque as a Flash Gordon movie for demonstrating the idea but it provides the basis for the engineering feats that put a man on the moon. The graphical presentation in this section will not be picturesque either. However, it does provide the basis for an engineering view of economics. I provide the "trajectories" of the strategies; how they behave over time. Though I provide only models or examples here, it is important to realize that the significance of these examples is that a quantitative description is possible; just like a spaceship with the correct initial conditions you can compute the exact trajectory of the ship: The dynamic theory of games provides a deterministic prescription for how the odds change for the mixed strategies of economic game theory.

I caution however that the trajectories require some interpretation; they are not the movie. The problems in economic theory are harder than the problems of spaceship flight. The dimension of the strategic space in economics is much higher than the physical space of three dimensions and most of us are not adept at visualizations in such higher dimension spaces. Because of the difficulties of visualization, at best I will be able to hint at the actual picture, even though the trajectories in fact provide full access to the total picture.

Before coming to the graphical presentation, I deal with a few preliminaries. The mathematical analysis of Chapter 5 suggests that specific real world games might have certain inactive choices. For numerical analysis, I make a specific though arbitrary choice and take a zero-sum model in which there are either two or three players, each with exactly two strategies. I assume that time is an isometry in the central frame.

For the two-person zero-sum game, I assume that the angular deviations are also isometries. I choose the player form to reproduce a given constant market field f_{ab}. The form of the line element is given by Eq. (5.4). This line element has three remaining unknown functions: g_{00}, p and A_0. I focus on the behaviors in the central frame. I consider games with both zero and non-zero game values. I pick representative forms for the unknown functions, assuming them to depend only on the active radial strategies.

One consequence of looking at "real" trajectories is they force issues that might be passed over in the analytic discussion. In particular, the quantitative calculations force clarity when discussing strategies and computing payoffs. Payoffs are the means the theory has for stating the amount a player will receive from any other player at a specific point in time. The normal concept of payoff forces the following distinctions. For each ***strategy–choice*** x^m, (the trajectories in economic theory) which must be non-negative[35], I assume there is a ***strategy scalar*** y^m that can be both positive and negative, such that

$$y^m = x^m - \left(x^m\right)^{-1}. \tag{6.1}$$

I assume the form Eq. (5.4) is given in terms of the player strategy scalars. I define the ***strategy–flow*** of the player as the appropriately transformed flow:

$$\frac{dx^a}{d\tau} = \frac{\partial x^a}{\partial y^b}\frac{dy^b}{d\tau}. \tag{6.2}$$

I compute the strategy–choice by inverting the defining relationship:

$$x^m = \tfrac{1}{2}\left(y^m + \sqrt{\left(y^m\right)^2 + 4}\right).$$

In this way I assure that the strategy–choice is non-negative. For large strategy–choices, the two are the same. The payoff F_{ab} is defined in the frame of the strategy scalar coordinates. To obtain the payoff between two players in the frame of the strategy–choices, I multiply the payoff F_{ab} by the transformation for each player:

$$\frac{\partial y^m}{\partial x^n} = \delta^m_n\left(1 + \left(x^m\right)^{-2}\right). \tag{6.3}$$

I also multiply the payoff by the normalized strategy–choice:

$$\frac{x^m}{\sum_{n \in player\ \alpha} x^n}.$$

---

[35] It may be possible to relax the concept of strategy–choice and choices in general to be both positive and negative. In this case one might also define a covariant payoff in terms of the market field and flows, which will be independent of the coordinates chosen.

The sum is over the strategy–choices of the player. The game value is the payoff from player α to β and is determined in terms of the payoff matrix F_{ab}:

$$payoff_{\alpha\beta}(y) = \sum_{\substack{m \in player\,\alpha \\ n \in player\,\beta}} \frac{x^m + \left(x^m\right)^{-1}}{\sum_{m' \in player\,\alpha} x^{m'}} F_{mn}(y) \frac{x^n + \left(x^n\right)^{-1}}{\sum_{n' \in player\,\beta} x^{n'}}. \quad (6.4)$$

The (non-covariant) expression is not singular (none of the denominators can vanish) and the structure that appears at small strategy–choices disappears when the strategy–choices are sufficiently large. Though not elegant, for the numerical exercises I think this is a reasonable interpolation function with which to discuss the payoffs as commonly defined.

6.1 Fair Games

The simplest examples are fair games, games in which each player receives at most a zero payoff. In this case I take both strategic lengths r_1, r_2 to be inactive, as well as time in the *central* frame and the angular deviation strategy θ_ω defined in the central frame. A consequence is that the time–choice in the *player* frame is also inactive. The gravitational potential and pressure are functions of a single active variable, the radial deviation strategy r_ω associated with the fair game.

6.1.1 No gravity or pressure

I start with the gravitational potential $\nu \equiv \frac{1}{2}\ln g_{00}$ and the pressure set to zero and a representative ***strategic mass*** or ***time–scale*** $m_0 = 4$. For the fair game examples the payoff matrix is chosen to be:

$$G = \begin{pmatrix} -2 & 1 \\ 4 & -2 \end{pmatrix}.$$

Before proceeding with the numerical analysis however, I provide an example for what this payoff matrix might represent. From *The Art of War*, Sun Tzu (1988), it is well understood that a larger army prevails over a smaller army, at least when the smaller army fails to take into account the size of its combatant. I consider therefore a game with two armies with slightly different strategies. The generals know that in general a large army (strong) will win against small army (weak). One side (the smart side) however has assessed its strengths and so changes how it fights using a small army by adopting a guerilla strategy of not fighting in the open (smart):

$$\left(\begin{array}{c|cc|c} \text{Payoff} & \text{strong} & \text{weak} & \\ \hline \text{strong} & 0 & 3 & 2/3 \\ \text{smart} & 6 & 0 & 1/3 \\ \hline & 1/3 & 2/3 & \end{array}\right).$$

So large beats small but smart makes small succeed over large. A small but smart army can pick off the enemy without much loss. The advantage disappears against a small force, since the other side can stay hidden as well. The expected payoff is 2 units to the army that plays smart. I have made the game fair by subtracting this amount from all payoffs, which doesn 't change the strategic behavior of either side.

The strategic behavior of the smart side should be to use its large army $2/3$ of the time and its guerilla army $1/3$ of the time. The behavior of the other side is to use its strong army $1/3$ of the time and its small army $2/3$ of the time. In this way each army adjust to the capabilities of the other side. For example if the smart side always fought with its guerilla force, then the other side would realize this and always fight using a small army. So it pays for the smart side to use its large army some of the time.

The behaviors I have described are the expected limiting case. A dynamic theory is meant to describe the dynamic changes that might occur to the odds played. These are given by the trajectories of the strategy–choices Fig. 6.1 or equivalently their flows as shown in Fig. 6.2.

In this model, all strategy–choices are approximately linear with path length and hence in time. The visualization of this in strategic space is that of a point moving with constant velocity. Like the bets at a roulette table, the strategy–choices stack up on squares representing the strategy choice of each player; the longer the game is played, the more the tokens are stacked up.

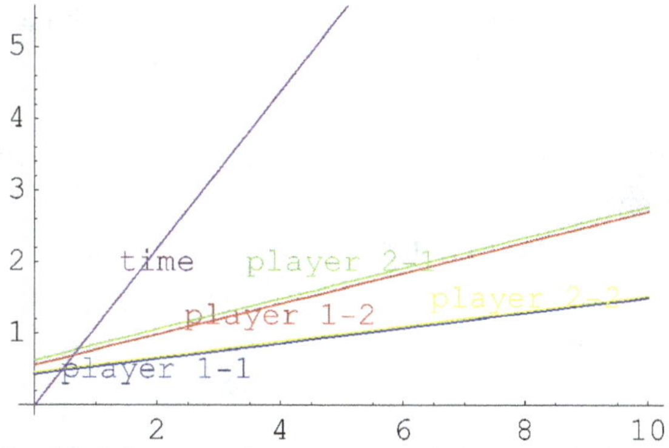

Fig. 6.1 Fair-game player strategy–choices (vertical axis) versus path length (horizontal axis) with no pressure or gravitational field—Player two strategy–choices are blue and red, player one strategy–choices are green and yellow and time is purple.

The flows, Fig. 6.2, are approximately constant in path length and also in time. In addition to the constant component, there is also a cork-screw component.

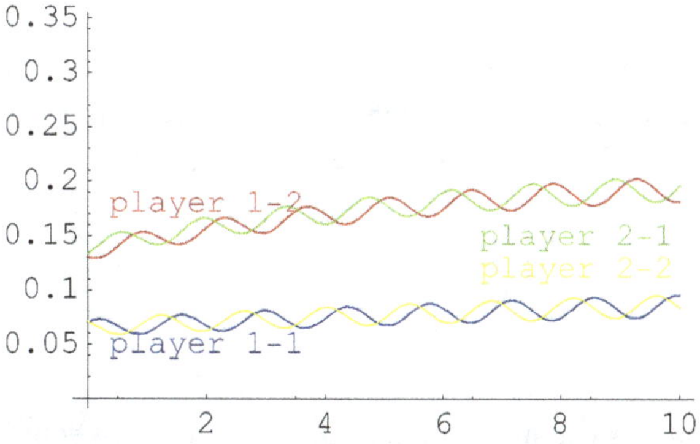

Fig. 6.2 Fair-game player strategy–flows with no pressure or gravitational field—Player two strategy–choices are blue and red, player one strategy–choices are green and yellow and time is off scale.

This clearly reflects a dynamic mechanism. The strategic lengths are by construction linear in path length. In this example, the angular player deviation is also by construction linear in path length whereas the radial player deviation is constant. If the two players start with an equilibrium flow, there will be no structure. This is because the initial angular deviation rate is zero. I have started the model with a small angular deviation rate and it is clear from the small cork-screw behavior of the figure that it remains.

The behavior of the trajectories of the strategy–choices determines the value of the payoff matrix, Fig. 6.3 and the player odds, Fig. 6.4. The player payoff approaches the equilibrium value zero. The economic content of the game is specified by this value.

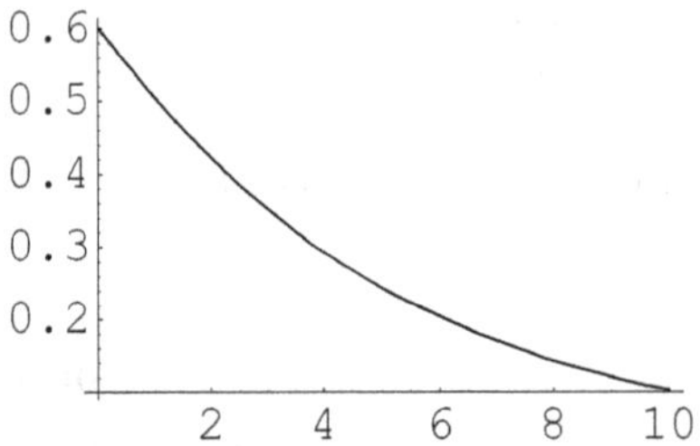

Fig. 6.3 Fair-game payoff or value (vertical axis) versus path length (horizontal axis) presented for the case with no pressure or gravitational field.

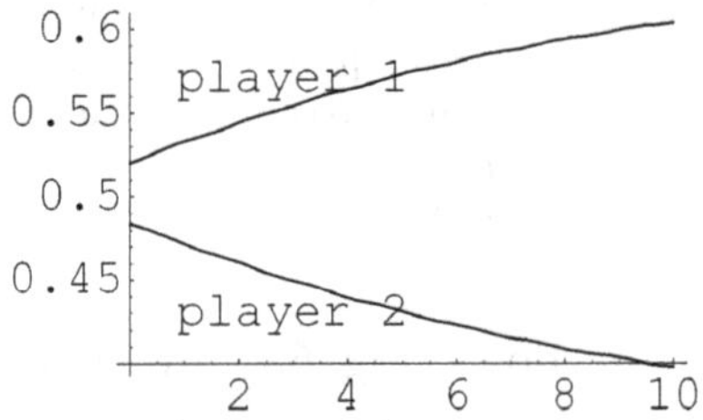

Fig. 6.4 Fair-game normalized player strategy–choices (vertical axis) versus path length (horizontal axis).

In the example of the two armies, though I start each army playing their strategies with near equal probability, they each migrate towards the appropriate equilibrium value. Based on Fig. 6.1 they do this with substructure and in a deterministic way that is set by the equations and initial conditions.

6.1.2 Gravity

In the previous example, if I start the game with an equilibrium flow, the flow remains at equilibrium no matter where I start the model in choice–space. I now consider an example where there are forces, in this case gravitational, that pull the model away from equilibrium except at a single fixed point in space. This example[36] has a non-zero gravitational potential $\nu \equiv \frac{1}{2}\ln g_{00}$ modeled in terms of the metric:

$$\nu = \frac{\nu_0}{1 + r_\omega^{\ 2}/r_0^{\ 2}}.$$

The field is zero at large radial deviation strategy distances and $\nu = \nu_0 \leq 0$ at the origin.

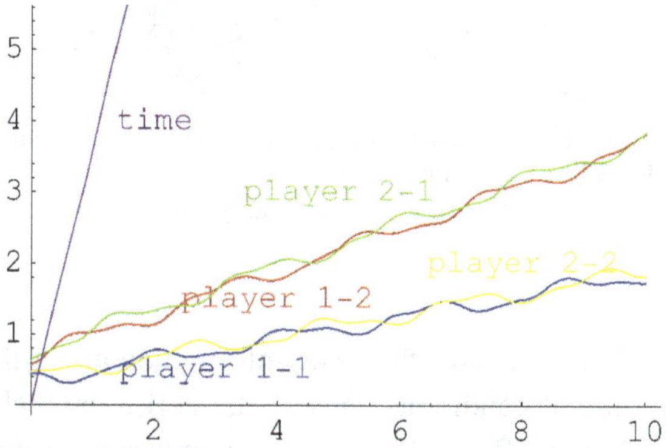

Fig. 6.5 Fair-game player strategy–choices (vertical axis) versus path length (horizontal axis) with a gravitational field—Player two strategy–choices are blue and red, player one strategy–choices are green and yellow and time is purple.

The visualization Fig. 6.5 of the strategy–choices is that of an object moving with an almost constant speed but whose path looks like a corkscrew. Because this occurs in a five dimensional space, it is not possible to draw this but the behavior of the components of the strategy–choices reflects this behavior: The strategy–choices are not strictly linear in path length but demonstrate fluctuations representing the corkscrew trajectory. On the roulette table, the players not only add to each pile but also remove tokens from the pile. This reflects a type of uncertainty in their behavior.

---

[36] The model parameters are: $\nu_0 = \ln\sqrt{\frac{1}{10}}$, $r_0 = \sqrt{\frac{1}{10}}$, the strategic mass is $m_0 = 8$ and the payoff matrix is $G = \begin{pmatrix} -2 & 1 \\ 4 & -2 \end{pmatrix}$.

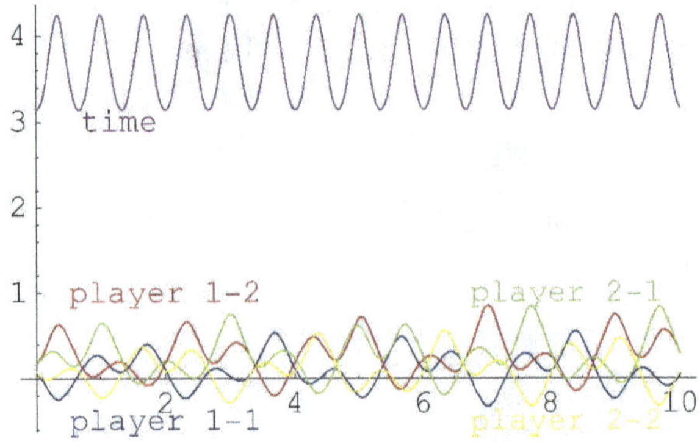

Fig. 6.6 Fair-game player strategy–flows with a gravitational field—Player two strategy–choices are blue and red, player one strategy–choices are green and yellow and time is purple.

Strictly speaking, we have the basic information about the trajectory. Because of the initial conditions chosen, the corkscrew behavior of the trajectory is a small effect. There are a number of ways to see the behavior more clearly. In the analogy with a spaceship, the speed variations might show a corkscrew trajectory more clearly than the trajectory itself. In our economic theory I have called these the strategy flows along each pure strategy. The flows are shown in Fig. 6.6. They are not constants but clearly oscillate, demonstrating the basic behavior Eq. (5.13) and its generalization of the assumption that the angular deviation is inactive.

First I see that time–flow is not constant but itself demonstrates a 10% structural change along the path. The player strategy–flows also demonstrate structure. They oscillate around an average value that represents the average flow (speed) for that pure strategy. The effect of the corkscrew behavior is evident in the correlations between the speeds. For example, I see that when player two puts more chits in one strategy he puts less in the other. The same holds for player one. This reflects this model's assumption that the total rate of strategy–choices for each player is constant.

It might be relatively clear that the trajectory is a corkscrew but the cause of the effect is not necessarily apparent. Of course in one sense, the cause resides in the equations, which give rise to the curves. The question however is to identify the source of the oscillation in a way that teaches us what to expect in the general case. By construction, this model has only a gravitational effect in addition to the game matrix. The gravitational potential is no longer zero or constant along a path, which is a reflection of the square of the player deviation difference as shown in Fig. 6.7.

The radial deviation strategy is a known function of the strategy-choices, hence already determined by Fig. 6.5:

$$r_\omega^{\ 2} = \bar{\bar{y}}^a \bar{\bar{\tau}}(\omega)_{ab}\, \bar{\bar{g}}^{bc} \bar{\bar{\tau}}(\omega)_{cd}\, \bar{\bar{y}}^d = y^a \tau(\omega)_{ab}\, g^{bc} \tau(\omega)_{cd}\, y^d .$$

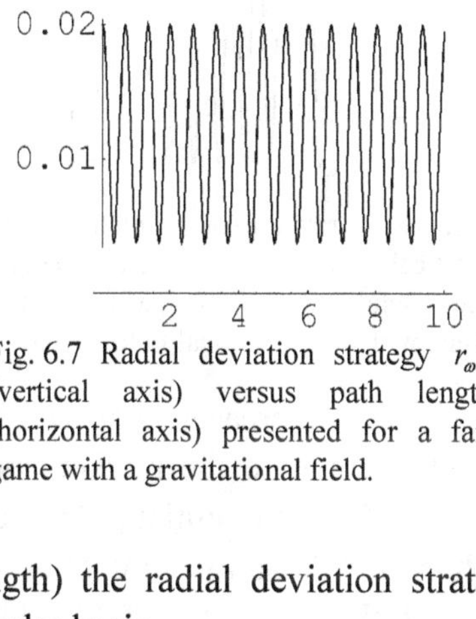

Fig. 6.7 Radial deviation strategy $r_\omega^{\ 2}$ (vertical axis) versus path length (horizontal axis) presented for a fair game with a gravitational field.

If the trajectory were strictly along the equilibrium straight-line path in choice–space, this square $r_\omega^{\ 2}$ would be zero. Because of the corkscrew trajectory however, this square varies with time. Because of the assumption that the effect of gravity is represented by a function only of $r_\omega^{\ 2}$, I expect gravity to be strongest when the radial deviation strategy is zero. In Fig. 6.7, I see that over time (path length) the radial deviation strategy moves from large to small on a regular basis.

Consider the spaceship analogy: If it were circling the earth at a constant distance above the earth, then it would constantly see the same strength of gravity. If however the ship were in an oscillating orbit that went sometimes closer and sometimes further away, then at the closer points the ship would see a stronger gravitational field than when it was farther away. At the closer points, it would have accelerated and would be moving slightly faster. The acceleration is indicative of a positive force. At the further points, it would decelerate and be moving slightly slower. The deceleration is indicative of a negative force. Do we see the same thing in this economic example, even though the dimensionality of space is much larger? I first compute the stationary market boost Fig. 6.8

(as seen in the central frame) as a function of path length from the flow components.

The market boost in Fig. 6.8 is a measure of the "speed" of the game. I note that the market boost varies in path length and hence in time. At the start (zero path length) the speed is at its lowest value of about 0 and increases to its maximum value of about 0.6. Because it increases I say there is a positive or attractive force causing the behavior. This pattern is repeated about 15 times on the graph.

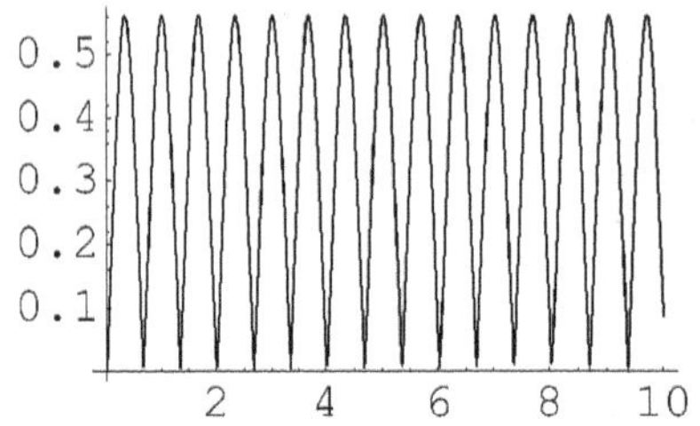

Fig. 6.8 Stationary market–boost (vertical axis) versus path length (horizontal axis) presented for a fair game with a gravitational field.

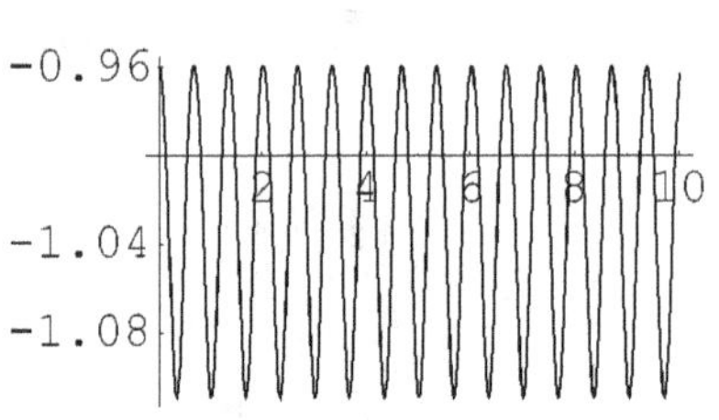

Fig. 6.9 Gravitational potential ν (vertical axis) versus path length (horizontal axis) presented for a fair game with a gravitational field.

To see the expected behavior that the system increases in speed as it moves toward a larger gravitational field, I compare the market boost Fig. 6.7 with the gravitational potential Fig. 6.9. At the initial point (zero path length), the speed is smallest and the radial deviation strategy $r_\omega^{\ 2}$ is largest (0.02). This corresponds to the smallest value of the gravitational potential ν. Conversely when the speed is largest (0.6), the radial deviation strategy $r_\omega^{\ 2}$ is smallest (0.005), corresponding to the deepest penetration into the gravitational well.

I get an interpretation that is more general than the example: The system falls into the gravity well, increasing its stationary market boost. The gravitational well is an attractive force. The above graphical result is a reflection of a general result based on the fact that in these models, time is an isometry and so there is a conserved charge Eq. (5.6). I recall that the approximate form of this equation is:

$$\tfrac{1}{2}\beta^2 + \nu + \phi + \frac{c_{\xi^0}}{n_0} A_0 \cong \text{constant}. \tag{6.5}$$

The correlated behavior described is a consequence of this isometry. The economic behavior is determined again by the value of the game shown in Fig. 6.10 and the strategies (*i.e.* the normalized strategy–choices) shown in Fig. 6.11. The numerical computation suggests that the fixed point of the solution is at the equilibrium value.

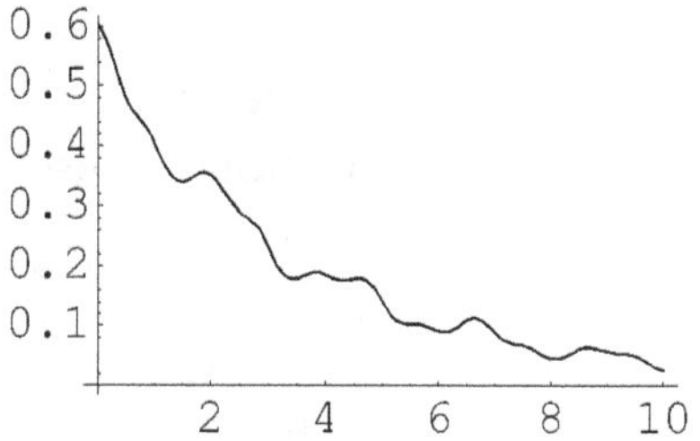

Fig. 6.10 Game value (vertical axis) versus path length (horizontal axis) presented for a fair game with a gravitational field.

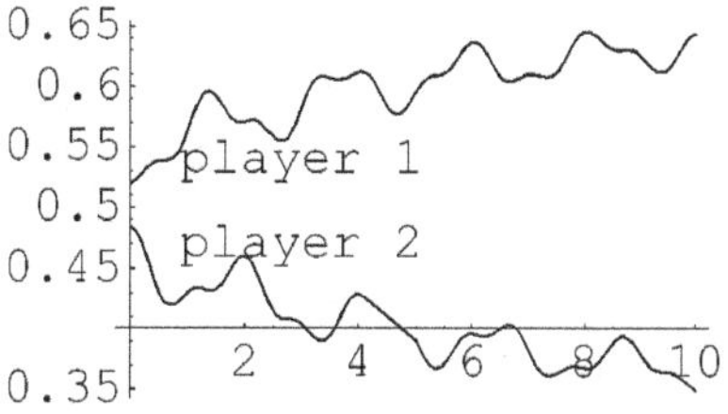

Fig. 6.11 Fair-game normalized player strategy–choices (vertical axis) versus path length (horizontal axis) with a gravitational field.

The normalized strategy–choices shown in Fig. 6.11 approach their equilibrium values, though with considerably more structure than the last example. The strategy–choices reflect a type of player uncertainty about which strategy to choose. This uncertainty is a prediction of the dynamic theory and should be subject to quantitative comparison with realistic economic behavior. This is engineering, not just a statement of qualitative behavior.

The theory provides more than just the individual behaviors but a statement about their correlations shown in Fig. 6.12. I see a corkscrew semi-periodic behavior of the solutions which goes from left to right and the periods are related to the interplay of the market and gravitational forces.

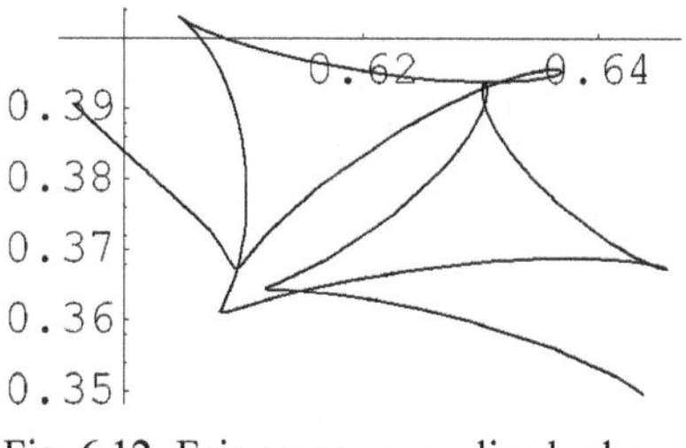

Fig. 6.12 Fair-game normalized player strategy–choice correlations between player two (vertical axis) and player one (horizontal axis) with a gravitational field.

In the example of the two armies, the "smart" general plays

along the horizontal axis and the other general along the vertical. From the starting point in the calculation, the two generals proceed towards their equilibrium as a consequence of the gravitational force attempting to minimize the radial deviation strategy. At a certain point however the speed of the game becomes too great and pushes the system away from the equilibrium. Over time, the two players play at their equilibrium values. The structure is what you should expect to see as the game is played multiple times.

With this example, I have so far demonstrated the trajectory and the economic consequences of the model. The results follow from a subset of the field equations. The example also provides some insight into the other field equations and what they must do for consistency. Because the stationary market boost is not zero and there is non-zero flow, the model with gravity also generates market currents Fig. 6.13 from the market field, a flow of "stuff" that in turn should generate additional market fields [*Cf.* Eq. (4.3)].

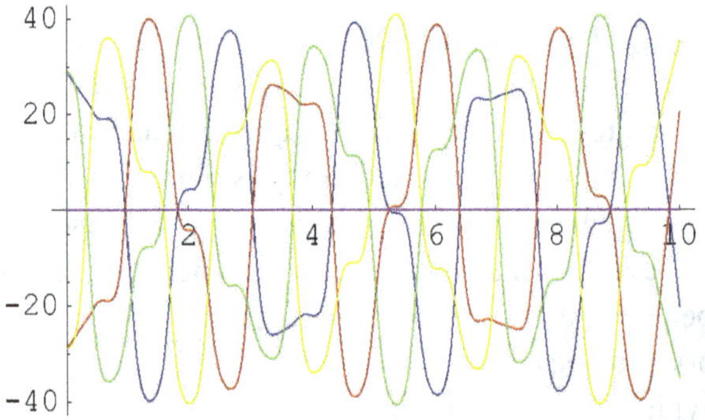

Fig. 6.13 Fair-game market currents (vertical axis) versus path length (horizontal axis) with a gravitational field—Player two currents are blue and red, player one currents are green and yellow and time current is purple.

An attribute of the solution is that the time component, J_0, the total strategic length components for player two $J_1 + J_2$ and player one $J_3 + J_4$ are zero. In particular the time component of the flow is the charge density, whose source depends on the payoff matrix components F_{0m}. These components are zero at equilibrium for a fair game. I obtain a generalization of this for the dynamic model: The source for these matrix components vanishes. In general however there will be sources that generate non-trivial behavior for the other market components. I conclude that a constant market field is an approximation to the full set of field equations.

6.1.3 Pressure

I now consider the effect of pressure by introducing a non-zero pressure field. In the last example, I learned that Eq. (6.5) was useful in determining what to expect with gravity. I use this to guess what the effect of pressure should be. For zero gravity, I see that a decrease in pressure should be balanced by an increase in speed. The movement from high pressure to low pressure reflects an attractive force as indicated by the increase in speed. In weather [Crawford (1992)] this is the notion that wind flows from high to low pressure, with a force proportional to the pressure gradient. As the wind moves down a pressure gradient, its speed increases.

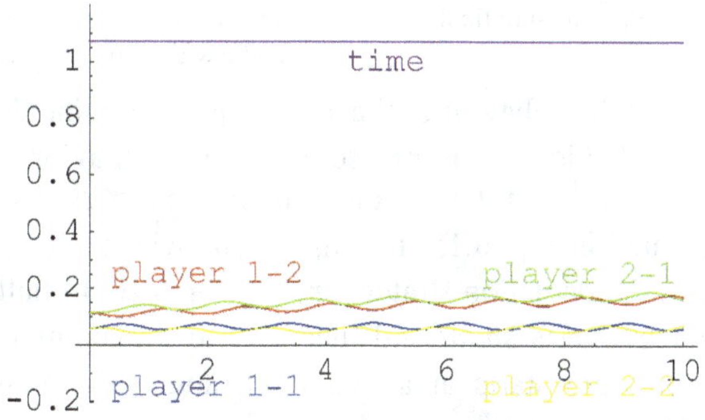

Fig. 6.14 Fair-game player strategy–flows (vertical axis) versus path length (horizontal axis) with a pressure field—Player two strategy–flows are blue and red, player one strategy–flows are green and yellow and time is purple.

In this example[37] I assume no gravity well and a pressure $p = \exp\left(-br_\omega^{\ 2}\right)$ that drops off as a function of the radial deviation strategy squared, $r_\omega^{\ 2}$. I don't show the trajectories in this case as they correspond to a path that is almost straight but again displays a slight corkscrew behavior around a straight-line direction. I show the flows in Fig. 6.14, which better illustrate this.

There is still player uncertainty about which strategy to choose. Each flow component is oscillatory around its average value. Time is almost constant as a function of path length. The strategy–choices for each pure strategy therefore increase with time, modified by the small corkscrew like behavior.

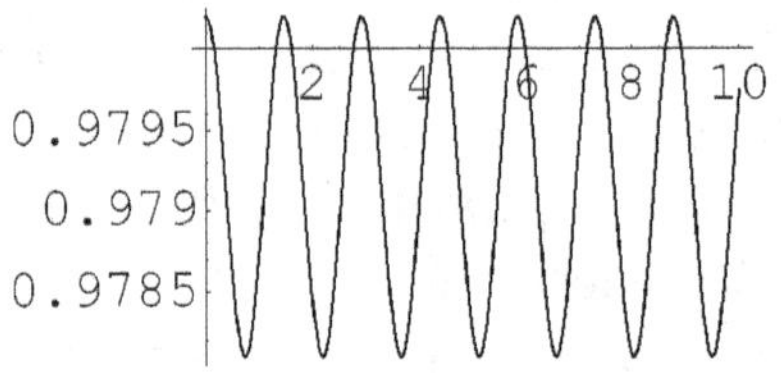

Fig. 6.15 Pressure (vertical axis) versus path length (horizontal axis) presented for a fair game with a pressure field.

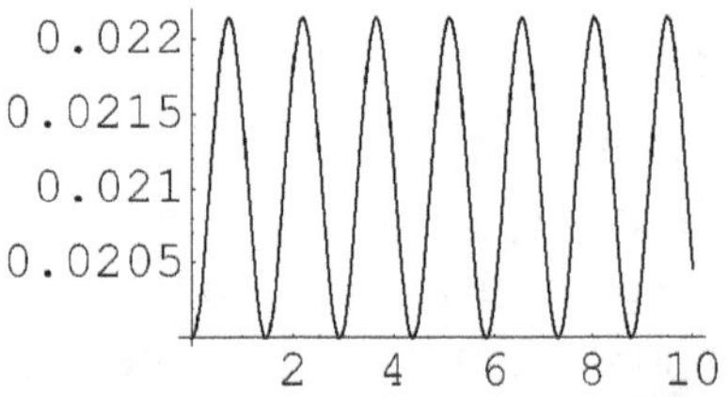

Fig. 6.16 Stationary radial deviation strategy $r_\omega^{\ 2}$ (vertical axis) versus path length (horizontal axis) presented for a fair game with a pressure field.

To understand the behavior of the system, at each point along the path I compare the behavior of the pressure with the behavior of the radial deviation strategy $r_\omega^{\ 2}$ and the stationary market boost β. I start with the pressure illustrated in Fig. 6.15. I compare this with the behavior of the square of the radial deviation strategy $r_\omega^{\ 2}$ in Fig. 6.16. Initially the radial deviation strategy starts small and increases to a maximum (Fig. 6.16) whereas the pressure starts at a maximum and goes to its minimum (Fig. 6.15). This correlation is a direct consequence of the model assumption $p = \exp\left(-br_\omega^{\ 2}\right)$. The interesting correlation is what happens to the stationary market boost shown in Fig. 6.17.

---

[37] The model parameters are: $b = 1$, $m_0 = 4$ and the payoff matrix is $G = \begin{pmatrix} -2 & 1 \\ 4 & -2 \end{pmatrix}$.

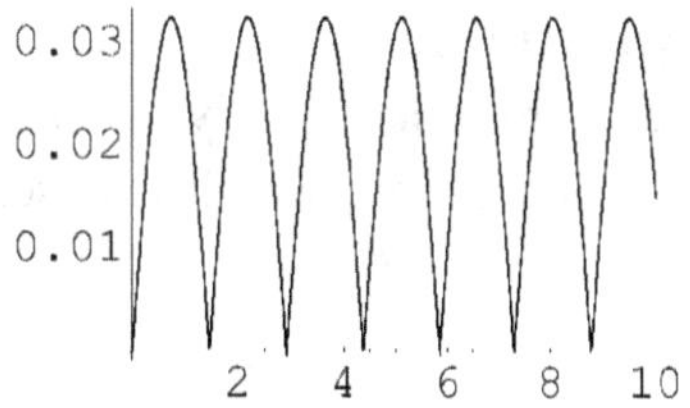

Fig. 6.17 Market boost (vertical axis) versus path length (horizontal axis) presented for a fair game with a pressure field.

As a function of path length, initially the stationary market boost increases (as the pressure decreases) until it achieves a maximum; at which point it decreases (as the pressure increases). This is a reflection of the conservation law Eq. (6.5).

The economic behavior is reflected in the behavior of the game value (Fig. 6.18) and normalized strategy–choices (Fig. 6.19). The game value decreases toward the equilibrium fair game value of zero. The normalized strategy–choices approach their equilibrium value as well: It is difficult to see the corkscrew type behavior of the trajectory in these plots.

I can capture some effect by looking at the two dimensional correlation between the normalized player strategy–choices and comparing this example with the last example (with a gravitational field). The correlation progresses from left to right as shown in Fig. 6.20.

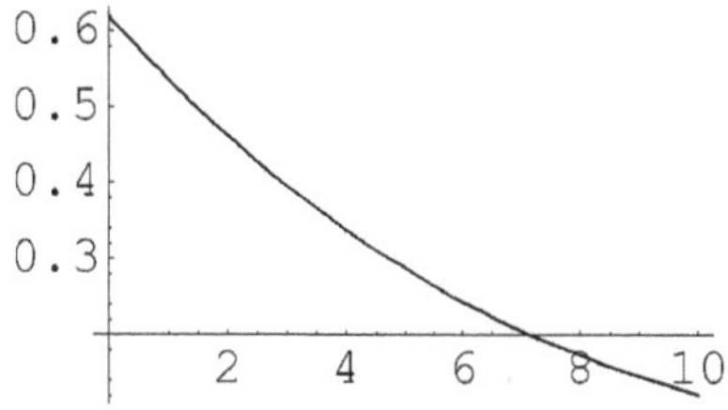

Fig. 6.18 Game value (vertical axis) versus path length (horizontal axis) presented for a fair game with a pressure field.

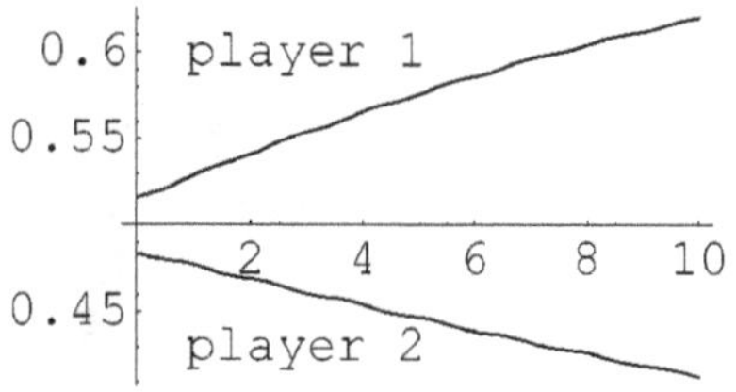

Fig. 6.19 Strategy–choices (vertical axis) versus path length (horizontal axis) with a pressure field.

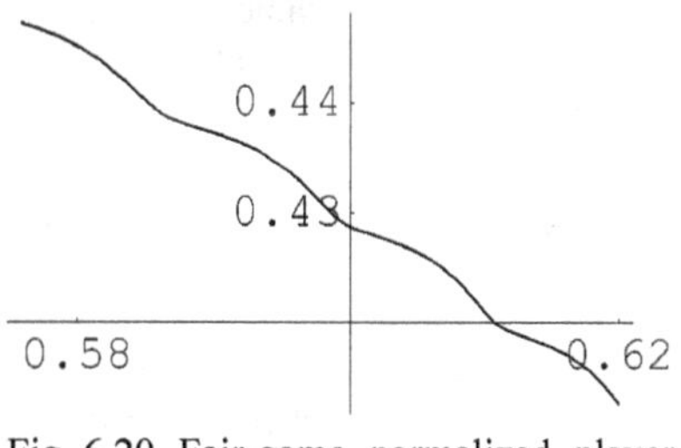

Fig. 6.20 Fair-game normalized player strategy–choice correlations between player two (vertical axis) and player one (horizontal axis) with a pressure field.

In the fair game examples, I have considered only some of the equations that govern the flow and have ignored the field equations that determine the gravitational field. These examples have only a single active strategy. For this case I can obtain a full solution, which I present in the next section.

6.1.4 Single strategy model

The fair games that I have described in the previous three examples have a single active strategy. In these examples I have made different assumptions about the parameters of the theory, parameters that characterize the strength of the gravitational potential, the pressure and the market bias. In principle all these models should fall in the category

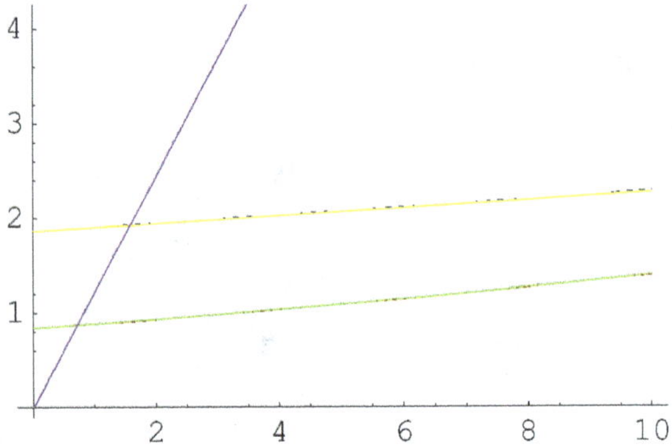

Fig. 6.21 Player strategy–choices (vertical axis) versus path length (horizontal axis) for the single strategy model— Player two strategy–choices are blue and red, player one strategy–choices are green (on top of red) and yellow (on top of blue) and time is purple.

of the single strategy model for which there is a full solution. In this section I provide an example of a model where the full solution to the field equations is used. I deal with the details of the model in Appendix E. I report here additional results including predictions for the normalized strategy–choices and the game value.

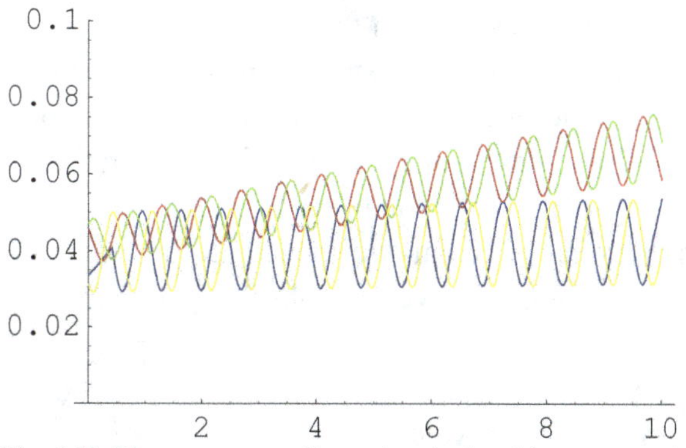

Fig. 6.22 Player strategy–flows (vertical axis) versus path length (horizontal axis) for the single strategy model—Player two strategy–choices are blue and red, player one strategy–choices are green and yellow and time is off scale.

The interpretation of the graphs will be similar to those from the previous examples.

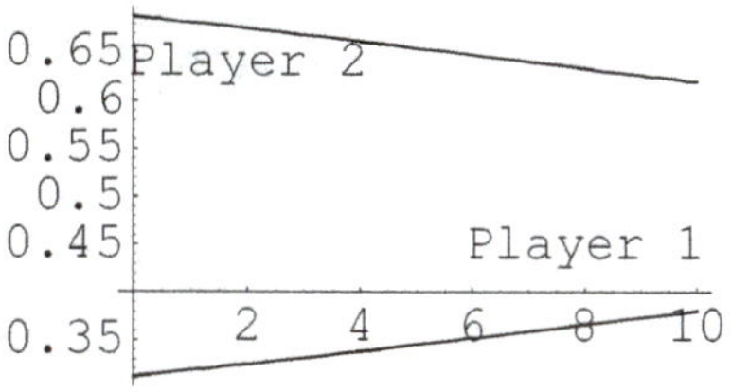

Fig. 6.23 Normalized player strategy–choices (vertical axis) versus path length (horizontal axis) for the single strategy model—solid line is player one and dashed line is player two.

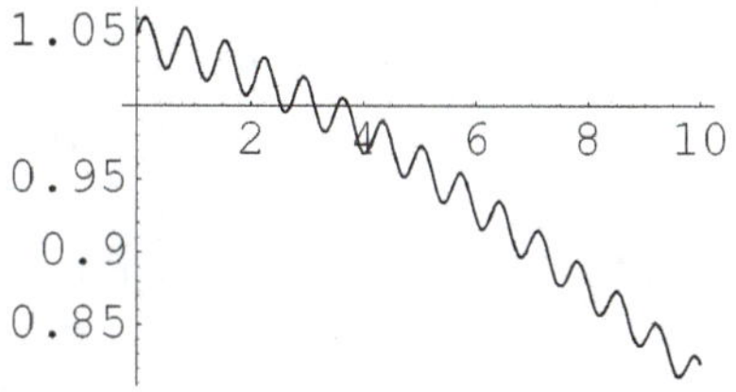

Fig. 6.24 Game value (vertical axis) versus path length (horizontal axis) presented for the single strategy model.

The normalized strategy–choices in Fig. 6.21 are all approximately linear. Though the forces have canceled, oscillations remain in the player frame because of the rotational symmetry. I see the oscillation structure in the strategy flows shown in Fig. 6.22.

The economics is correspondingly similar for the normalized strategy–choices shown in Fig. 6.23. The game value approaches zero as seen in Fig. 6.24. The full equations maintain the fair value of the game.

I see that the game value displays the rotational symmetry of the model by showing the fine structure from the oscillations.

The full solution to the single strategy model displays all the features of the initial example in Sec. 6.1.1. However both the forces, described in the examples with gravity and pressure, are in operation.

The solution to the full equations *requires* the pressure field (or control), gravitational potential and indeed all the metric elements to be constant along the path. However neighboring paths will be characterized by different values of these variables. The full equations determine these different values. In the following equations, I demonstrate the predictions of the full model as a function of the "natural length" u, which I identify with the radial deviation strategy. It is by definition a constant along the path and the only active strategy.

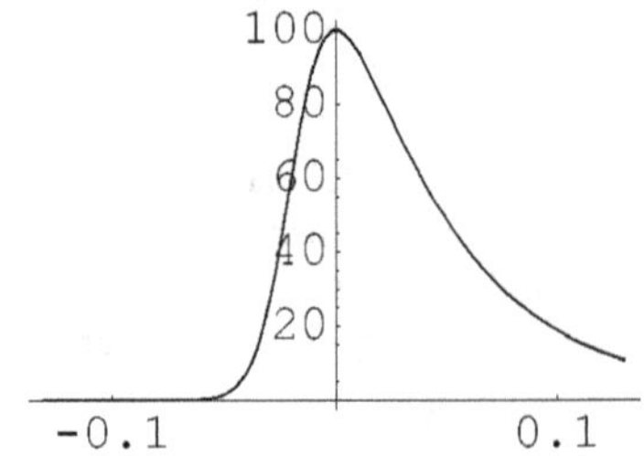

Fig. 6.25 Pressure (vertical axis) versus natural scale u (horizontal axis) presented for the single strategy model.

In Fig. 6.25 for the pressure, I see that there is a rapid drop-off away from the equilibrium position. As opposed to the reduced pressure, the full pressure has its maximum at the equilibrium radial deviation-distance.

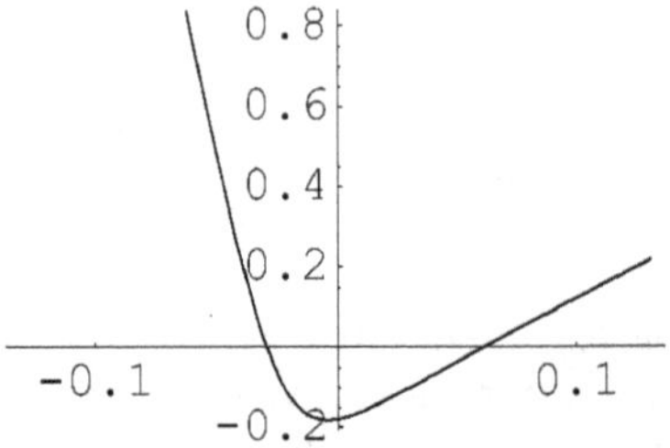

Fig. 6.26 Gravitational potential v (vertical axis) versus natural scale u (horizontal axis) presented for the single strategy model.

In Fig. 6.26 I see "proof" that there is a gravitational well. I see similar if not inverse structures between the pressure and gravitational field. Since each contributes to the force equation

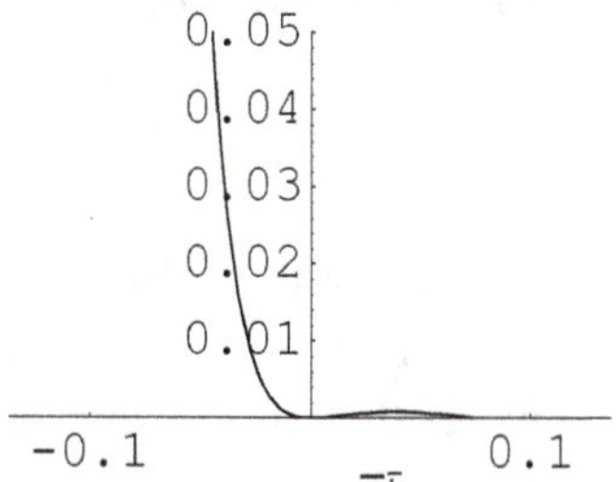

Fig. 6.27 Market bias $\overline{A}_0^k$ (vertical axis) versus natural scale u (horizontal axis) presented for the single strategy model.

in terms of their gradients, it should come as no surprise that these two fields produce effects which cancel. The single strategy model provides some motivation for the forms used in the previous models in the earlier sections. It is important to note that we have motivation from the dynamic theory, not analogy to other fields of study.

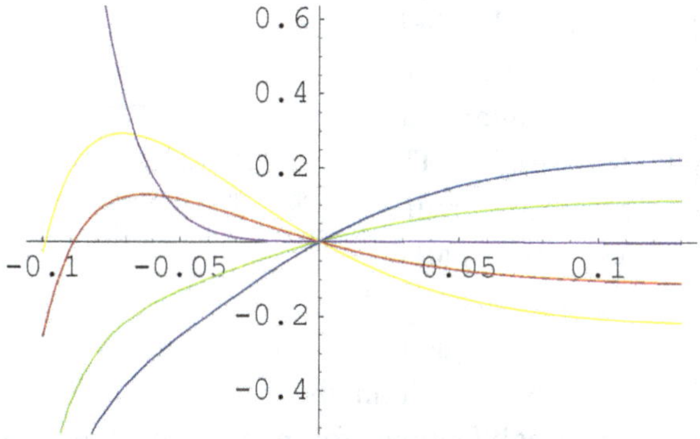

Fig. 6.28 Player one potential components (vertical axis) versus natural scale u (horizontal axis) for the single strategy model—Player two strategy–choices are blue and red, player one strategy–choices are green and yellow and time is purple.

In the previous examples I assumed a negligible ***market bias*** $\overline{A}_0^k$ as a function of the radial deviation strategy u associated with the time component of the market field in the player frame. In the single strategy model, it has a minimum at the equilibrium and is negligible for large distances. Does this suggest that at short distances self interest tradeoffs become more important? This form is suggestive of a form I might use for models in which solutions to the full equations are not available.

I note that the space components of the market field are generally not constants. I compute the market potential, from which the market field would be computed. I recall from Eq. (5.4) that I expect the market potentials to be linear in the strategic scalars:

$$\bar{A}^{\bar{k}}_{m(\alpha)} = -\tfrac{1}{2} f_{m(\alpha)n} y^n$$

I therefore expect such potentials in the player frame to be linear in the radial deviation strategy u. In Fig. 6.28 I see this is the case as an approximation around the equilibrium. I have shown the curves for player 1; the curves for player 2 are almost identical.

There are two consequences of note. First, the basic behavior of the field equations is to create a market field that is a multiple $\omega(r)$ of the original market field which is shown in Fig. 6.29. The game remains a fair game at all distance from the origin. The general behavior at large radial distances is that the market payoffs become small.

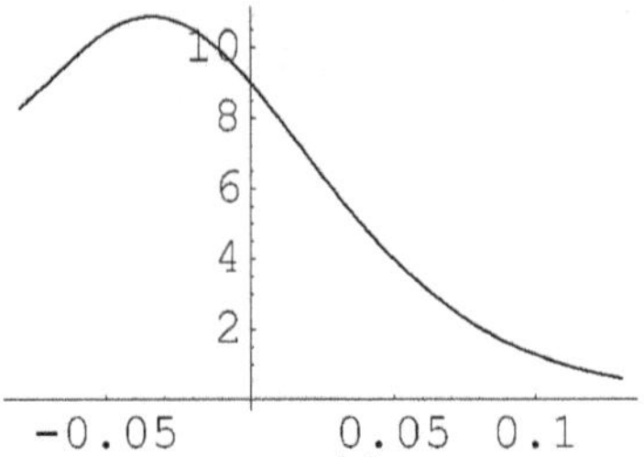

Fig. 6.29 Factor $\omega(r)$ (vertical axis) versus natural scale u (horizontal axis) presented for the single strategy model.

The second major consequence of note is that there are now internal factions. These are payoffs within the coalition called player one or player two. The internal factions for player one (vertical axis) as a function of the natural length (horizontal axis) is shown in Fig. 30.

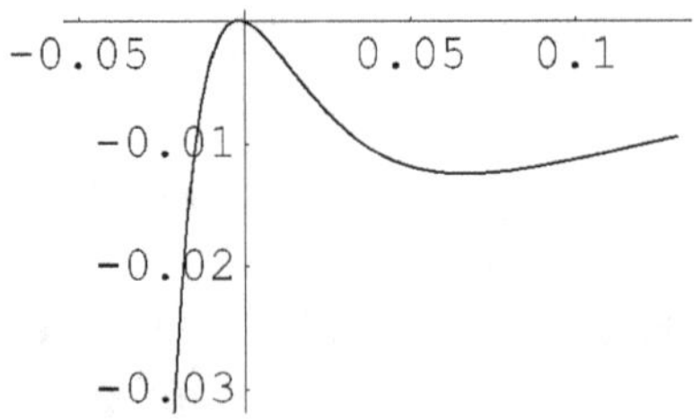

Fig. 6.30 Internal faction for player one (vertical axis) versus natural scale u (horizontal axis) presented for the single strategy model.

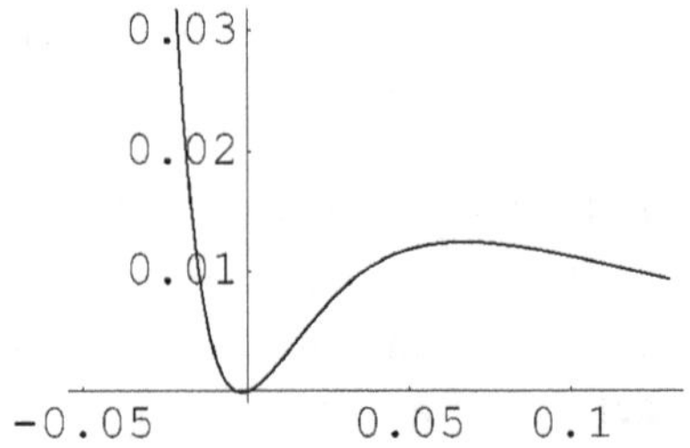

Fig. 6.31 Internal faction for player two (vertical axis) versus natural scale u (horizontal axis) presented for the single strategy model.

The corresponding result for player two is shown in Fig. 6.31. It is striking that the internal factions remain small at distances larger than equilibrium but become significant at distances shorter. It suggests that unusual behavior can be expected at short distances where there may be uncertainty about the game on the part of the players.

6.2 Value Games

So far, I have considered games that at equilibrium are fair and have essentially a single active strategy. They illustrate specific aspects of the theory. Though they are greatly simplified, they do reflect aspects of the single strategy model that has a complete solution. ***Value games*** in which at equilibrium one player receives more than any other player, introduce a new ingredient. In these games I suggest there are generally two or more active strategies. I expect the periodic motion seen in the previous sections to become semi-periodic or indeed ***chaotic***. This is like the change from (closed) great circle routes on a sphere to chaotic geodesics on an oblate spheroid Fig. 2.1.

Value games are described in Sec. 5.2. For the class of models considered here in which each player has two possible pure strategies and where the two angular strategies $\{\theta_v, \ \theta_\omega\}$ are inactive, there will be two active strategies corresponding to the two radial strategies: $\{r_v, \ r_\omega\}$. Both strategies are forced to be active since in polar coordinates [Eq. (5.14)] the metric depends on both radial directions.

In the following examples, the active strategies are the two radial strategies in the central frame. The inactive strategies are the two angular strategies and the hidden player strategy. Time as measured in the central frame is also inactive. The new ingredient is a non-zero value component and the possibility of a non-trivial radial value coordinate dependence. One consequence is that even with the full equations, I no longer expect the active strategies to be constant along the path. There can now be a flow from one active strategy to the other[38].

---

[38] *Cf.* Eq. (C.1). With a single active strategy, the flow must vanish as there is no market field.

For the purpose of this analysis, the exact value of the game is not essential, so unless stated otherwise, I take examples with a game value of two and I use a variable ***market mass*** $m_0 = 4$ and a payoff matrix

$$G = \begin{pmatrix} 0 & 3 \\ 6 & 0 \end{pmatrix}.$$

This is the payoff for the example of two armies described in Sec. 6.1.1.

6.2.1 No gravity or pressure

The first example is with the market pressure and gravity absent. This example highlights the effect of the value component. If I start with equilibrium flow, the trajectory is along a constant direction in choice–space and there are no deviations away from this. The strategy flows are constants and time and all the strategy–choices increase linearly with path length. This is similar to the result for a fair game.

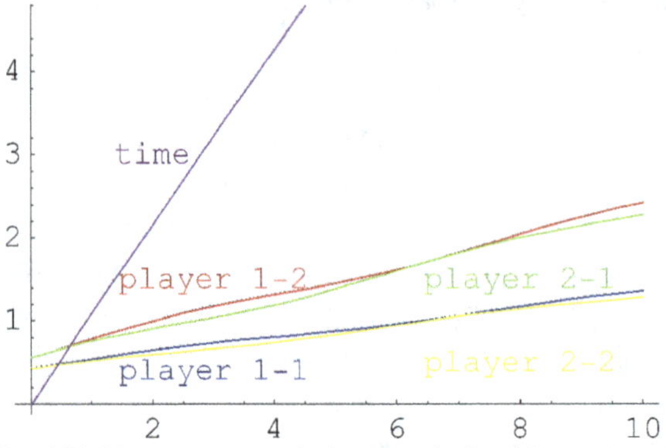

Fig. 6.32 Player strategy–choices (vertical axis) versus path length (horizontal axis) for a game with value and no pressure or gravitational fields—Player two strategy–choices are blue and red, player one strategy–choices are green and yellow and time is purple; all quantities evaluated in the player frame.

I start with a small deviation away from equilibrium flow and obtain a trajectory, Fig. 6.32, which is approximately along a line in space but also displays a corkscrew behavior with two distinct frequencies. The second frequency (compared to a fair game) is a consequence of the game value. As with all previous examples, I see that time flows positively and is approximately linear in path length.

Also, as in past examples, the corkscrew behavior is seen more clearly with the strategy flows, the speed along each of the pure strategy directions. The strategy–flows oscillate around their equilibrium values, displaying two frequencies showing the generalization to the basic behavior Eq. (5.13). This is seen clearly in Fig. 6.33.

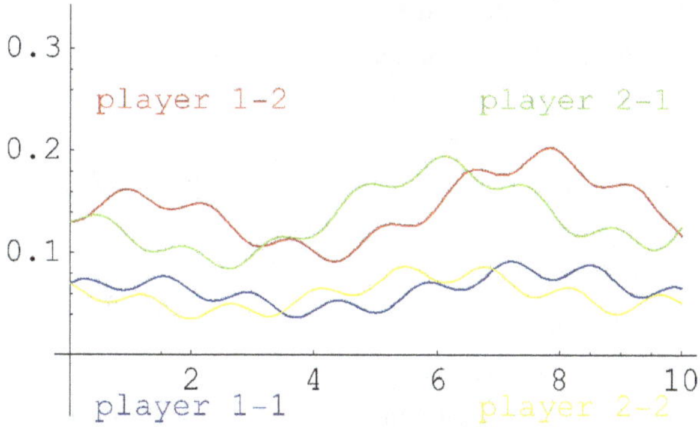

Fig. 6.33 Player strategy–flows (vertical axis) versus path length (horizontal axis) for a game with value and no pressure or gravitational fields—Player two strategy–choices are blue and red, player one strategy–choices are green and yellow and time is off scale; all quantities evaluated in the player frame.

Each player displays some uncertainty as to which choice is supported. On a roulette table, at the outset player two increases strategy–choices for both strategies while player one decreases strategy–choices for both strategies. The sum of strategy–choices for each player is not constant and shows a more complex behavior than in the fair game examples. During the course of many plays of the game, there are times when player one seems to show more confidence than player two (player one places strategy–choices on both strategies whereas player two removes strategy–choices from both strategies) and then this is reversed.

I note that if the number of strategies to each player increases, the number of frequencies possible also increases. The structure becomes yet more complicated. In choosing my examples I have purposely limited the complexity in order to highlight some cause and effects.

The game value is the source of the new structure and reflects itself in the behavior of the time component of the market potential A_0, I name the ***market bias***. If the game starts at equilibrium flow, the market bias is zero and there are no forces pushing the system away from equilibrium. However, with the initial conditions chosen, the market bias starts near zero and oscillates around zero as seen in Fig. 6.34.

Again I call on Eq. (6.5) for help in the understanding. I see that in the absence of other forces, a linear combination of the market boost (squared) and the market bias are constant. In the central frame however, the market bias is zero and so the stationary market boost should be constant. Figure 6.35 shows that this is true within the numerical accuracy of the computation.

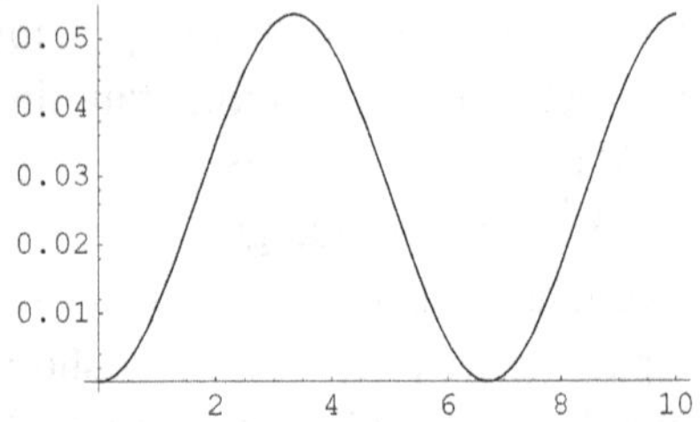

Fig. 6.34 Market bias (vertical axis) versus path length (horizontal axis) presented for a game with value and no pressure or gravitational fields.

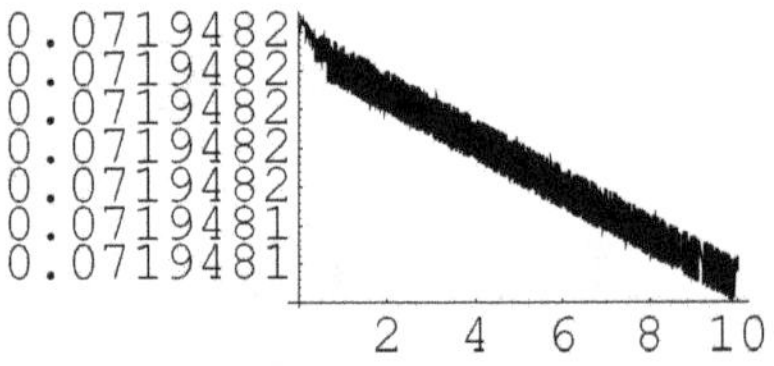

Fig. 6.35 Stationary market boost in the central frame (vertical axis) versus path length (horizontal axis) for a game with value and no pressure or gravitational fields.

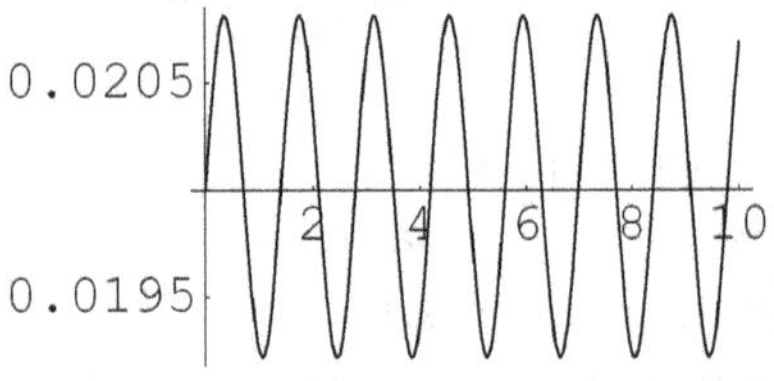

Fig. 6.36 Radial deviation strategy squared r_ω^2 (vertical axis) versus path length (horizontal axis) for a game with value and no pressure or gravitational fields.

Both radial strategies are bounded. The radial value strategy grows (Fig. 6.36) and only the radial deviation strategy is bounded (Fig. 6.37) with a different characteristic frequency in contrast to the fair game examples..

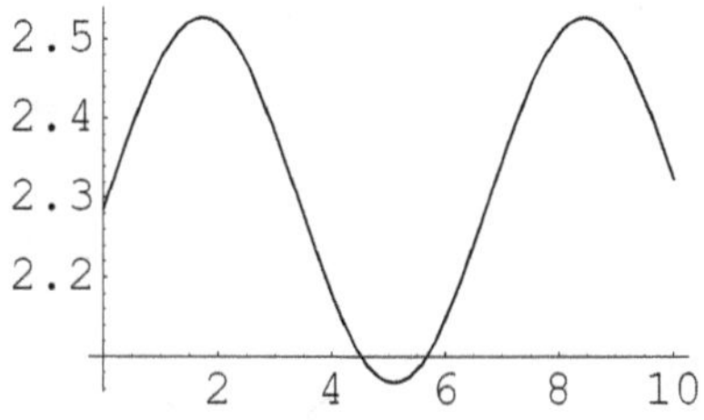

Fig. 6.37 Radial value strategy squared r_v^2 (vertical axis) versus path length (horizontal axis) for a game with value and no pressure or gravitational fields.

I thus have a way to demonstrate in a multi-dimensional space that there is truly a corkscrew type behavior. The successive plays of the game attempt to stay near the equilibrium flow.

I now turn to the economic aspects of the game as represented by the game value (Fig. 6.38) and the normalized strategy–choices (Fig. 6.39). The game value falls toward the equilibrium value. The behavior is dominated by the accumulated strategy–choices and only slightly influenced by the fluctuations around the average behavior. The fluctuations are small, corresponding to the small deviation in flow made from the defensive position.

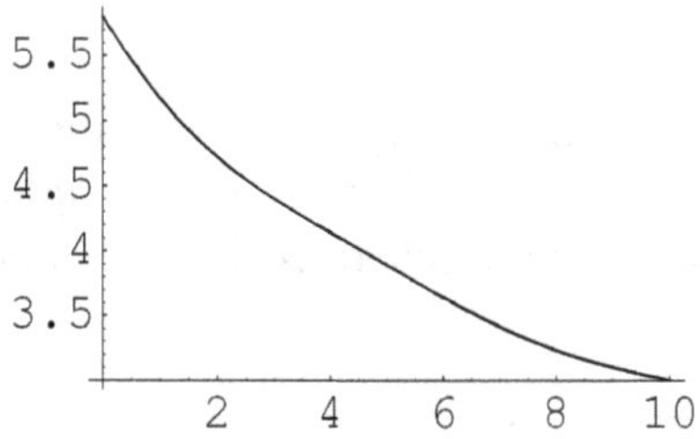

Fig. 6.38 Game value (vertical axis) versus path length (horizontal axis) presented for a game with value and no pressure or gravitational fields.

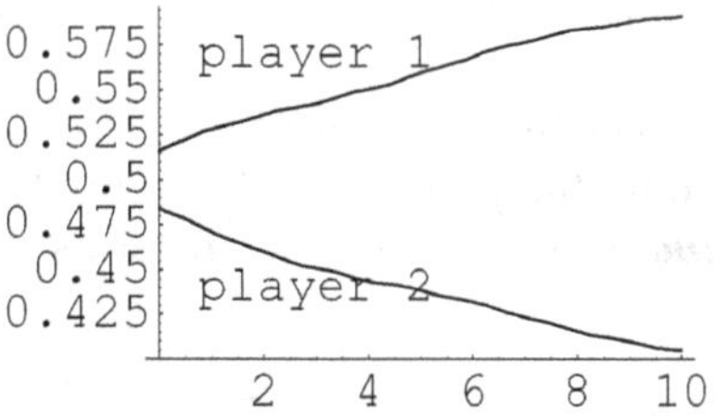

Fig. 6.39 Normalized player strategy–choices (vertical axis) versus path length (horizontal axis) for a game with value and no pressure or gravitational fields.

Not only does the game value converge now to the equilibrium value but the normalized player strategy–choices converge to their equilibrium values.

Although the game value and normalized player strategy–choices reflect little structure, the structure of the corkscrew behavior of the trajectory is reflected in the correlation Fig. 6.40.

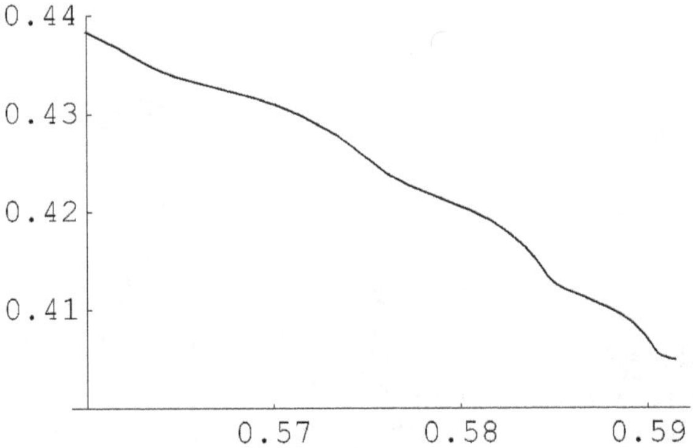

Fig. 6.40 Normalized player strategy–choice correlations between player two (vertical axis) and player one (horizontal axis) for a game with value and no pressure or gravitational fields.

It is interesting that there is correlation but neither previous corkscrew behavior observed for a fair game is present.

6.2.2 Gravity

I consider next a model with a gravitational field but no pressure. I assume the gravitational potential $\nu \equiv \tfrac{1}{2}\ln g_{00}$ is a function of the ***radial game strategy***, defined in terms of the rotation generated by the game matrix:

$$r_f^{\ 2} \equiv y^a f_{ab} g^{bc} f_{cd} y^d = \mathrm{v}^2 r_\nu^{\ 2} + \omega^2 r_\omega^{\ 2}.$$

By its definition, it is a function of both radial strategies and is a natural generalization of the fair game model. I assume the same form for the gravitational potential as I did for the fair game:

$$\nu = \frac{\nu_0}{1 + r_f^{\ 2} / r_0^{\ 2}}.$$

By this choice, I model a concept that the behavioral forces of the game depend on the game value as well as the underlying fair game. I also impose an isometry on the metric that it is unchanged by angular translations. These will be reflected in conserved charges.

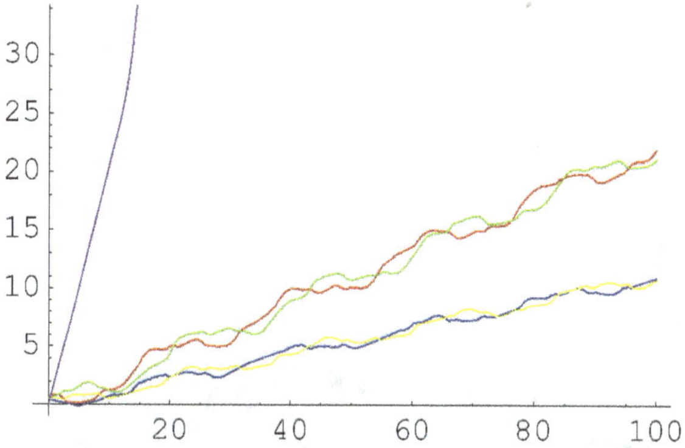

Fig. 6.41 Player strategy–choices (vertical axis) versus path length (horizontal axis) for a game with value, a gravitational field and no pressure field—Player two strategy–choices are blue and red, player one strategy–choices are green and yellow and time is purple; all quantities evaluated in the player frame.

In the following example, I start the system at the equilibrium flow but with values that generate large fluctuations[39]. This does not reflect periodic behavior but ***chaotic*** behavior.

The game strategy–choices demonstrate structure shown in Fig. 6.41. Time increases with path length; it is acceptable to think of the horizontal axis as proportional to time. The space components reflect a corkscrew-like behavior more complicated than the previous example, in part because the behavior is now far from equilibrium.

In the strategy flows, the ***chaotic*** structure shown in Fig. 6.42 is clear. The flows reflect the uncertainty of each player's choice. I have characterized the behavior as that of a corkscrew but I see that such a behavior is not sufficiently descriptive. There are two characteristic

---

[39] The other model parameters are: $v_0 = \ln\sqrt{1/10}$, $r_0 = \sqrt{8}$ and $m_0 = 8$.

frequencies at work. The behavior is quite different from the fair game Fig. 6.6, though on closer inspection, the underlying fair game structure provides the high frequency components. The low frequency components reflect a new structure leading to many possible corkscrews.

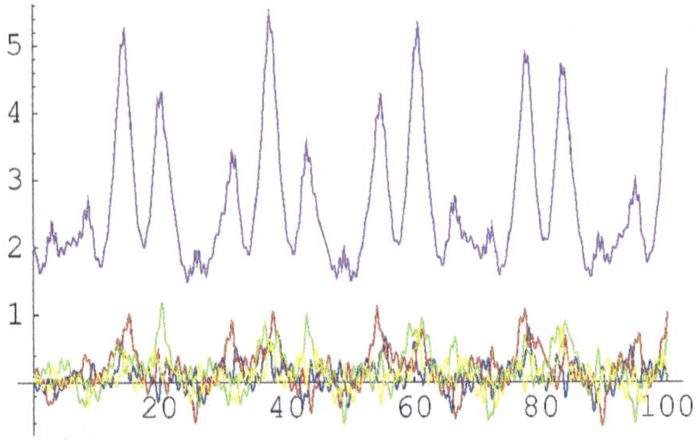

Fig. 6.42 Player strategy–flows (vertical axis) versus path length (horizontal axis) for a game with value, a gravitational field and no pressure field—Player two strategy–choices are blue and red, player one strategy–choices are green and yellow and time is purple; all quantities evaluated in the player frame.

Actually this should not come as a surprise. In three dimensions, a corkscrew is characterized by a preferred direction and a single rotation about the axis, which is the only possible rotation in the two dimensions transverse to the preferred direction. There is really only a left or right handed corkscrew. However, in four dimensions, I imagine again a single preferred direction but now there is a rotation group in three dimensions that defines the possible states transverse to that preferred direction. The rotation group in three dimensions is characterized by three directions, with three generators that exhaust all possible rotations. In our economic examples the space is five dimensional and the space transverse to the preferred direction is four dimensional. That suggests that the possibilities are characterized by rotations in four dimensions, which have six generators. The argument suggests that there might be six quite different looking corkscrews, of which we have identified only two. If nothing else, this argument helps clarify why it is difficult to visualize spaces of more than three dimensions. Certainly an open and interesting question will be to categorize the different corkscrew possibilities.

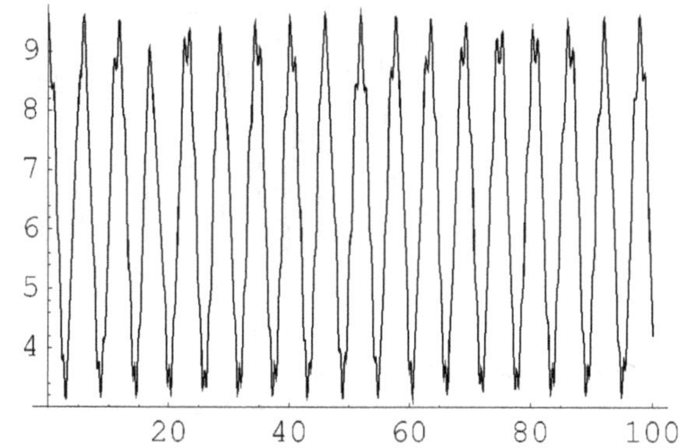

Fig. 6.43 Radial game strategy squared r_f^2 (vertical axis) versus path length (horizontal axis) for a game with value, a gravitational field and no pressure field.

The structure observed is in response to the gravitational force that is a function of the radial game strategy, Fig. 6.43. The high frequency components reflect the structure in r_ω^2 and the low frequency components reflect the structure in r_v^2.

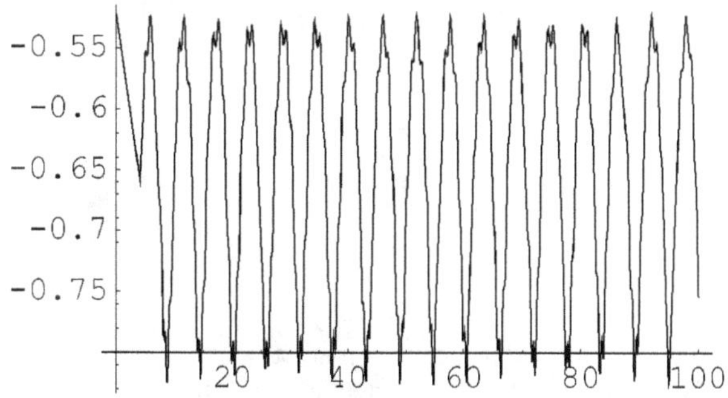

Fig. 6.44 Gravitational potential ν (vertical axis) versus path length (horizontal axis) presented for a game with value, a gravitational field and no pressure field.

The structure is reflected in the gravitational potential Fig. 6.44.

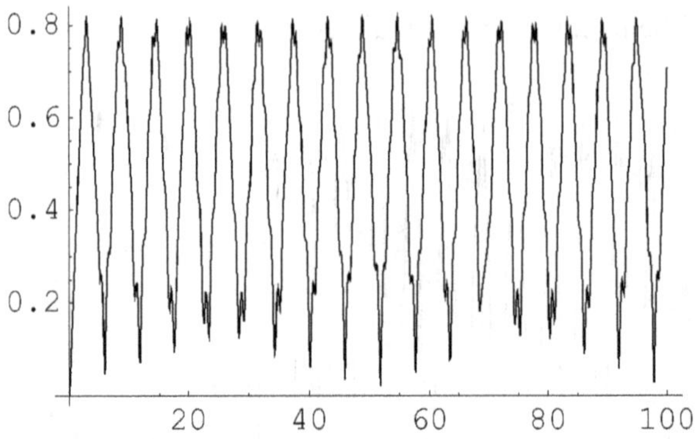

Fig. 6.45 Market boost (vertical axis) versus path length (horizontal axis) for a game with value, a gravitational field and no pressure field

The gravitational potential starts at a maximum for zero path length. As path length increases the gravitational potential drops. I compare this behavior with the stationary market boost Fig. 6.45. The speeds are very "relativistic".

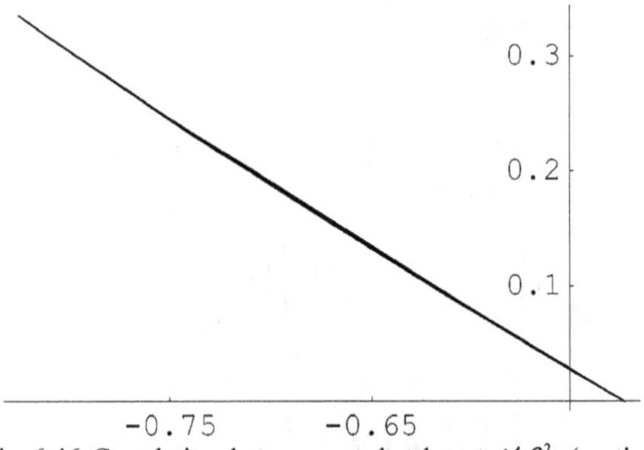

Fig. 6.46 Correlation between market boost $\frac{1}{2}\beta^2$ (vertical axis) and gravitational potential ν (horizontal axis) presented for a game with value, a gravitational field and no pressure field.

The behavior is complementary to that of the gravitational potential. This is seen in the scatter plot of the market boost and the gravitational potential, Fig. 6.46 that shows the expected linear correlation that follows from the conservation of energy.

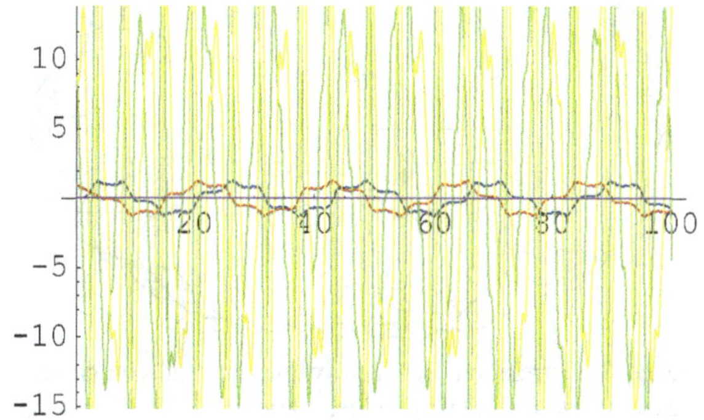

Fig. 6.47 Central frame market currents (vertical axis) versus path length (horizontal axis) for a game with value, a gravitational field and no pressure field—Player one strategic length current is blue, player two strategic length current is red, player one deviation currents is green, player two deviation current is yellow and time current is purple.

As with fair games, there can be currents generated by the divergence of the market field as shown in Eq. (4.3) resulting in Fig. 6.47. There are no currents along the time direction, an artifact of the model assumption of no market bias field in the central frame. In the (exact) single strategy model, this was not true away from the equilibrium position. The stationary currents associated with the strategic length of each player are small although the stationary flows are large. The deviation currents are much larger though their stationary flows are small. A full solution of the equations requires that the currents generated by the market fields balance the currents as calculated from the flows.

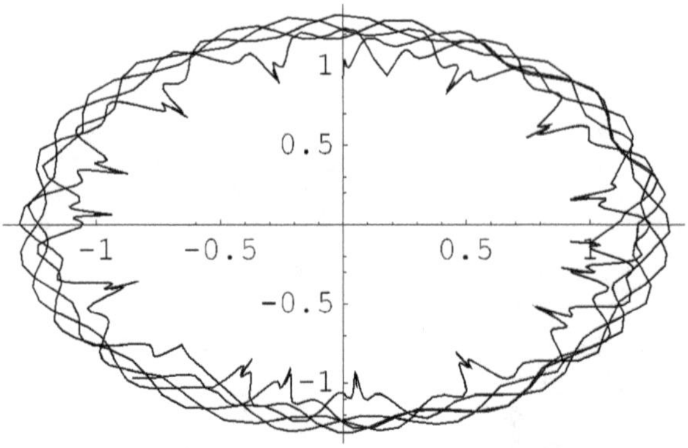

Fig. 6.48 Correlation between value current components presented for a game with value, a gravitational field and no pressure field.

In contrast to the fair game solutions, the stationary currents display a striking correlation when taken in appropriate pairs. I show first the central frame value correlations[40] in Fig. 6.48.

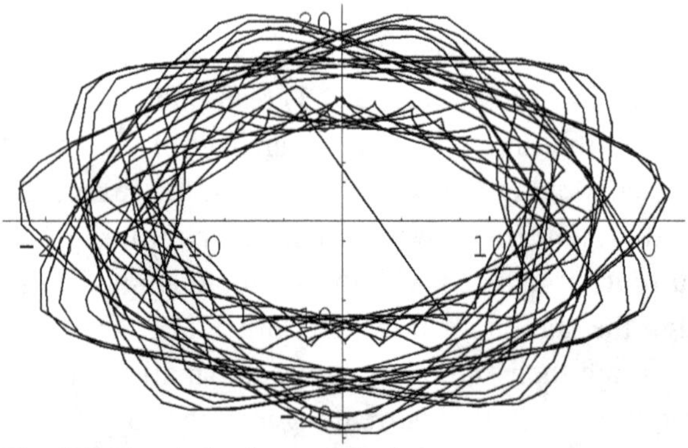

Fig. 6.49 Correlation between deviation current components presented for a game with value, a gravitational field and no pressure field.

I show next the central frame deviation correlations in Fig. 6.49.

[40] The next two figures illustrate the ***chaotic*** non-periodic nature of the solution.

Both currents reflect an oscillation of the orbits of the currents in a semi-periodic manner. This suggests that in the general case there will be flows from one strategy–choice to another.

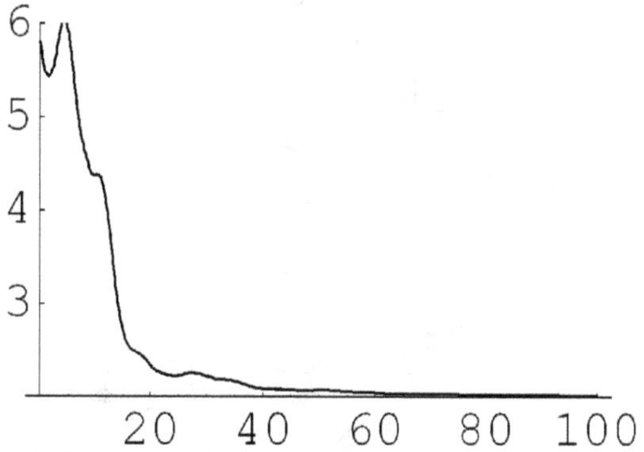

Fig. 6.50 Game value (vertical axis) versus path length (horizontal axis) presented for a game with value, a gravitational field and no pressure field.

The game value (Fig. 6.50) displays significant structure as it approaches its equilibrium value.

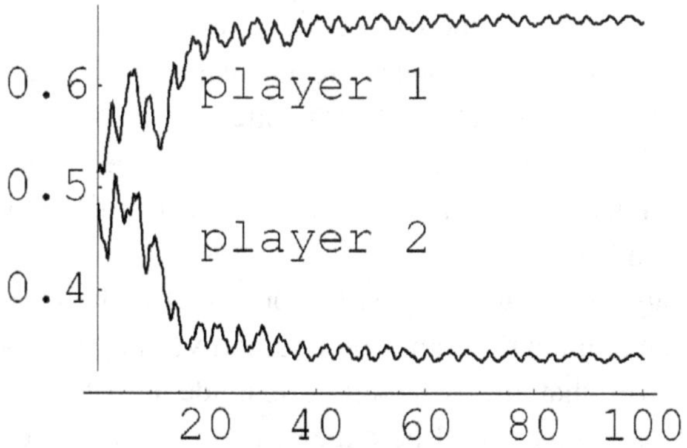

Fig. 6.51 Normalized player strategy–choices (vertical axis) versus path length (horizontal axis) for a game with value, a gravitational field and no pressure field.

The player strategy–choices (Fig. 6.51) display corresponding structure and also approach their equilibrium values.

The player strategy–choice correlations (Fig. 6.52) display a counter clockwise rotational behavior whose center moves to the right and down towards equilibrium. This has some similarity with the fair game of Fig. 6.12, fair-game correlations with gravity.

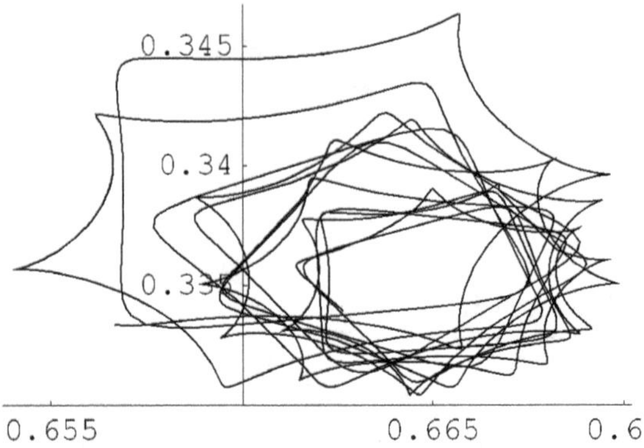

Fig. 6.52 Normalized player strategy–choice correlations between player two (vertical axis) and player one (horizontal axis for a game with value, a gravitational field and no pressure field.

I draw some tentative conclusions from this numerical exercise in the context of the example of the two armies. In the static theory of games, there are no strategic consequences to the value of the game. So if the two generals fought according to the game matrix of the fair game or according to the value game, they would play the same equilibrium strategy. The dynamic equations however reflect a difference. How are we to think of this difference?

One analogy that might help is to consider two bells of exactly the same shape and painted the same color so that visually they are the same. Imagine however that the two bells are made up of different metals. When struck, the bells would be expected to ring with a different note. We have that situation here.

In the two games, the generals appear to have the same choices (before the "bell" has been struck). The static situations are identical. The analogy of striking the bell is the action of starting each system at a point which does not correspond to equilibrium. What we see in the oscillation of the curves is the "ringing" of the non-linear system in response to a non-equilibrium start. By observing the differences, we in fact see that the "bells" are fundamentally different. The fact that a game has value makes its dynamic behavior fundamentally different from a fair game. I conclude that in contrast to the static theory, the game value has a strategic consequence.

What might a smart general learn from this? The purpose of a dynamic model is to obtain information about events that happen in time. The general may know that his strategy is good but not know how long it will take to train his troops to carry out that strategy. For his troops, there will be a learning process. In the case of a value game, the time constants are longer and so the learning process is longer. By understanding this and understanding the origin of the long time constants, a smart general might be able to modify them to fit his needs; he has controls or levers that he can now adjust. In this example, the controls are the strength and properties of the gravitational field. In the next subsection, I explore the response of the system with pressure to being "struck".

6.2.3 Pressure

Equation (5.7) admits a term due to the transverse pressure gradients. Pressure is a "political" influence on the game. It is not expected to force

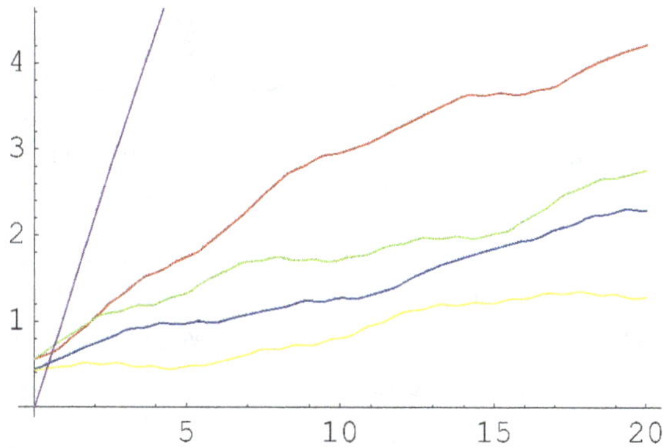

Fig. 6.53 Player strategy–choices (vertical axis) versus path length (horizontal axis) for a game with value, a pressure field and no gravitational field—Player two strategy–choices are blue and red, player one strategy–choices are green and yellow and time is purple; all quantities are evaluated in the player frame.

the players to their equilibrium position; in fact it may force them away. In the single strategy model, the pressure balanced the gravitational force. In a full solution, I expect both gravitational, bias and pressure forces. For now however I am looking to assess the interaction of the pressure gradient with the market force.

As with the previous example, I model the form of pressure to influence the behavior of the players. I assume the pressure depends on the radial game strategy $r_f^{\,2}$ as $p = \exp\left(-r_f^{\,2}/r_0^{\,2}\right)$. I start the model at rest when in the player frame. I show first the player strategy–choices[41] as a function of path length in Fig. 6.53.

With this choice of parameters, I have created a strong set of fluctuations, seen clearly in the strategy–flows in Fig. 6.54.

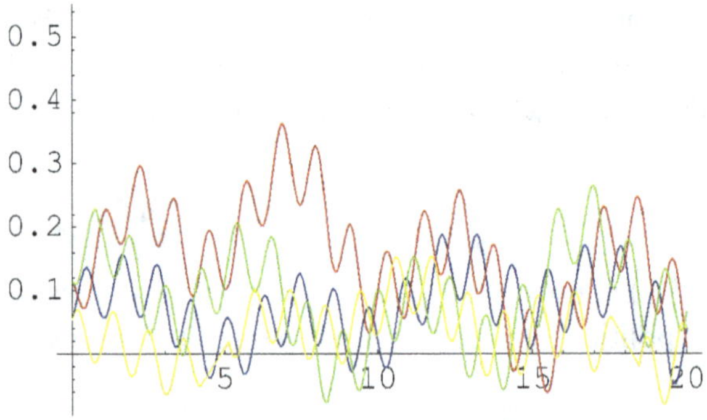

Fig. 6.54 Player strategy–flows (vertical axis) versus path length (horizontal axis) for a game with value, a pressure field and no gravitational field—Player two strategy–choices are blue and red, player one strategy–choices are green and yellow and time is off scale; all quantities are evaluated in the player frame.

There are two distinct oscillation periods as expected. The behaviors follow from the influence of the radial game strategy seen in Fig. 6.55.

---

[41] The value of the game is two and the other model parameters are: $r_0 = \sqrt{8}$ and $m_0 = 4$.

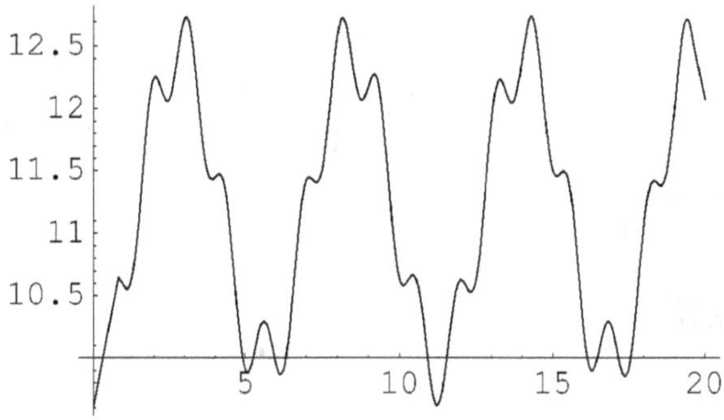

Fig. 6.55 Radial game strategy r_f^2 (vertical axis) versus path length (horizontal axis) presented for a game with value, a pressure field and no gravitational field.

This behavior determines the behavior of the pressure, Fig. 6.56.

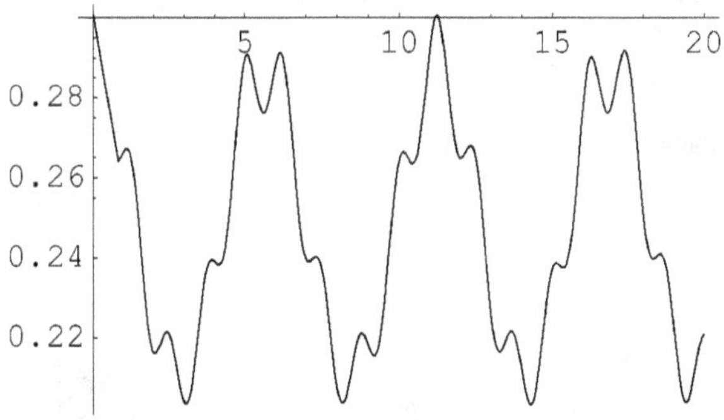

Fig. 6.56 Strategic pressure (vertical axis) versus path length (horizontal axis) presented for a game with value, a pressure field and no gravitational field.

It shows the pressure is initially a maximum and decreases initially. The overall behavior is oscillatory with two periods. The market boost, Fig. 6.57, increases initially, reflecting the effects of pressure that "blows a market wind" from high pressure to low pressure.

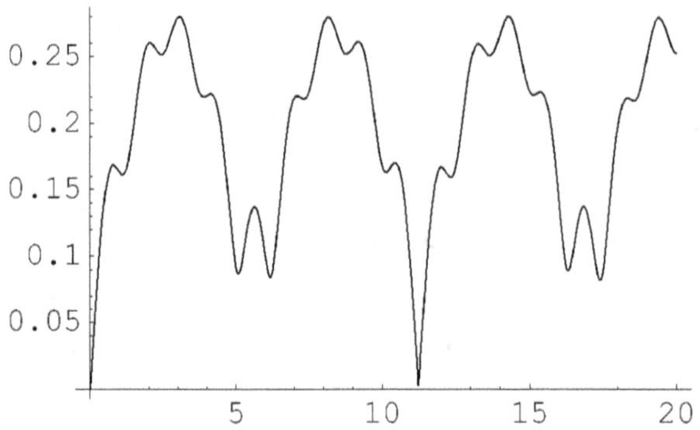

Fig. 6.57 Market boost (vertical axis) versus path length (horizontal axis) presented for a game with value, a pressure field and no gravitational field.

The oscillatory structure stays consistent with the time-charge conservation law as seen graphically by plotting the market boost and the pressure, Fig. 6.58; the expected (approximately) linear relationship results.

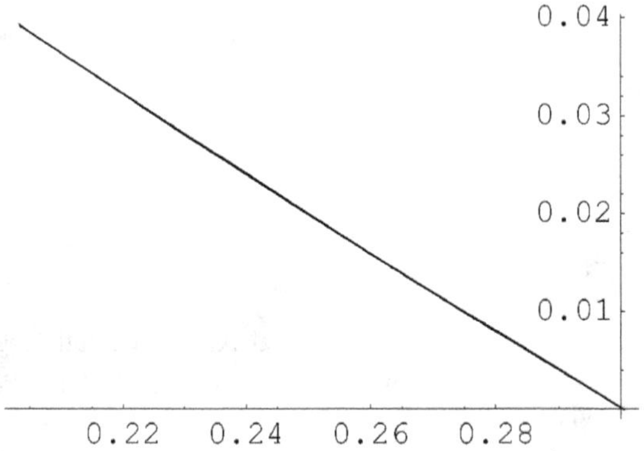

Fig. 6.58 Correlation between market boost $\frac{1}{2}\beta^2$ (vertical axis) and strategic pressure p (horizontal axis) presented for a game with value, a pressure field and no gravitational field.

The pressure does not prevent the market forces from moving the game value to equilibrium as seen in Fig. 6.59.

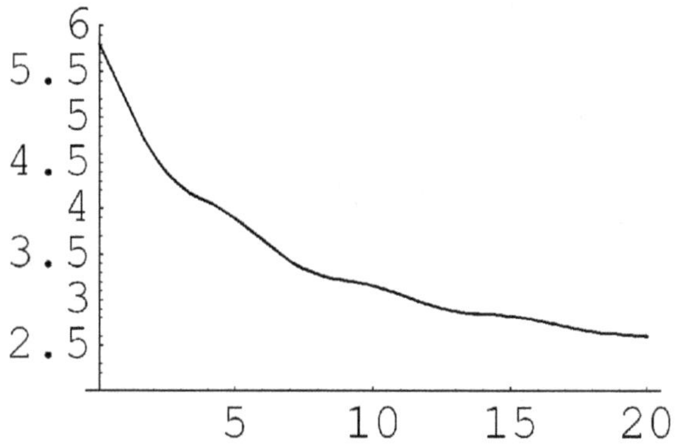

Fig. 6.59 Game value (vertical axis) versus path length (horizontal axis) presented for a game with value, a pressure field and no gravitational field.

The player strategies also approach their equilibrium values, as seen in Fig. 6.60.

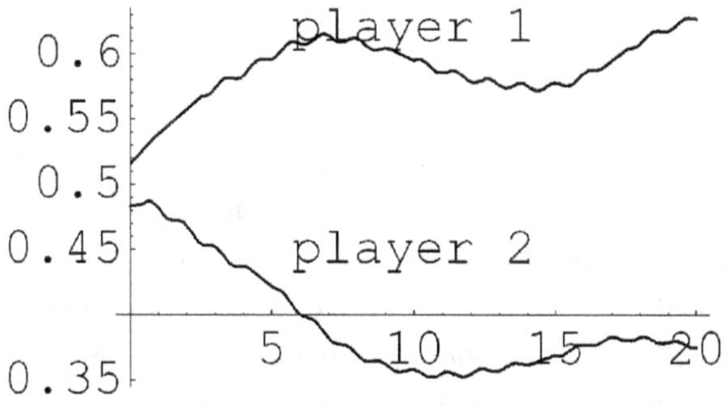

Fig. 6.60 Normalized player strategy–choices (vertical axis) versus path length (horizontal axis) for a game with value, a pressure field and no gravitational field.

The player strategy correlations move from left to right and rotate clockwise, as seen in Fig. 6.61.

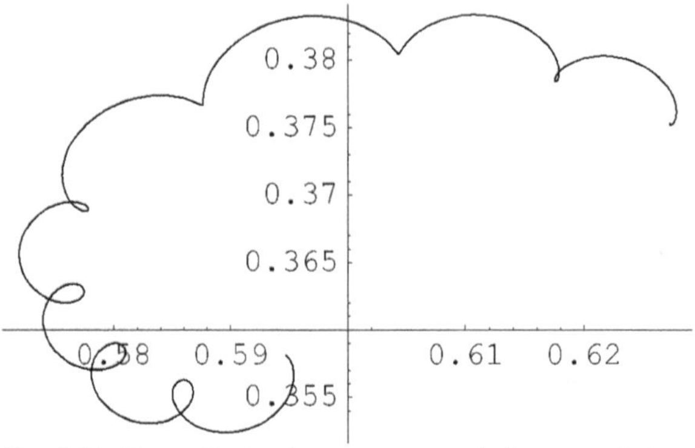

Fig. 6.61 Normalized player strategy–choice correlations between player two (vertical axis) and player one (horizontal axis for a game with value, a pressure field and no gravitational field.

The player correlations reflect the corkscrew type behavior, now with fine structure.

6.3 Three-Person Game

A three-person game has significantly more complexity. Any two players can form a coalition. Even with only two strategies per player, there will be more total strategies and hence the strategic space will be six dimensional. There are of course more possibilities for active strategies as well. Based on the insights gained from the previous section on two-person games, I consider an example in which the game matrix (the space components of the payoff matrix) is constant. The simplest choice is made for the components of the game matrix [Eq. (4.10)].

For the example below, the payoff to player one from players two and three and the payoff to player two from player three are as follows:

$$G_{12} = \begin{pmatrix} -3 & 0 \\ 3 & -3 \end{pmatrix} \quad G_{13} = \begin{pmatrix} -2 & 0 \\ 3 & -3 \end{pmatrix} \quad G_{23} = \begin{pmatrix} -1 & 0 \\ 3 & -2 \end{pmatrix}.$$

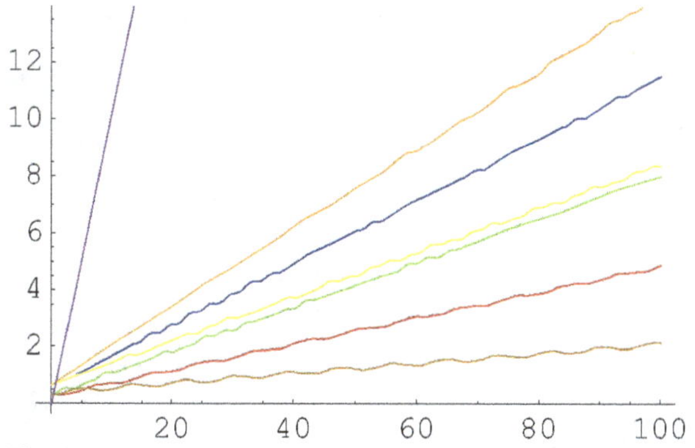

Fig. 6.62 Player strategy–choices (vertical axis) versus path length (horizontal axis) for a three person game—Player one strategy–choices are blue and red, player two strategy–choices are green and yellow, player three strategy–choices are brown and orange and time is purple.

The game matrix is formed from these payoffs and the assumption that the defensive position is in the null space of the market field. To compare with the static theory, I note that each player can play the game as if the other two players formed a coalition against him. There are three such two-person zero sum games. The respective payoff matrix for each player's defensive game follows from the above payoffs:

$$G_1 = \begin{pmatrix} -5 & -3 & -2 & 0 \\ 6 & 0 & 0 & -6 \end{pmatrix}$$

$$G_2 = \begin{pmatrix} 2 & 3 & -4 & -3 \\ 3 & -2 & 6 & 1 \end{pmatrix}$$

$$G_3 = \begin{pmatrix} 3 & -1 & -2 & -6 \\ 0 & 2 & 3 & 5 \end{pmatrix}$$

The value of each game is respectively $\{-2, \ -1/3, \ 1\}$. The sum of the values is $-4/3$, indicating that the game is not ***flat*** [for a discussion of flat games see Von Neumann and Morgenstern (1944)]: There is an advantage to each player to be in a coalition. Each player can receive at least the value of their defensive game. The amount they receive over and above that value is called the ***imputation***.

I start with the above model, assuming the central frame is the player frame, a strategic mass $m_0 = 20$ and compute the strategy–choices as a function of path length to obtain Fig. 6.62.

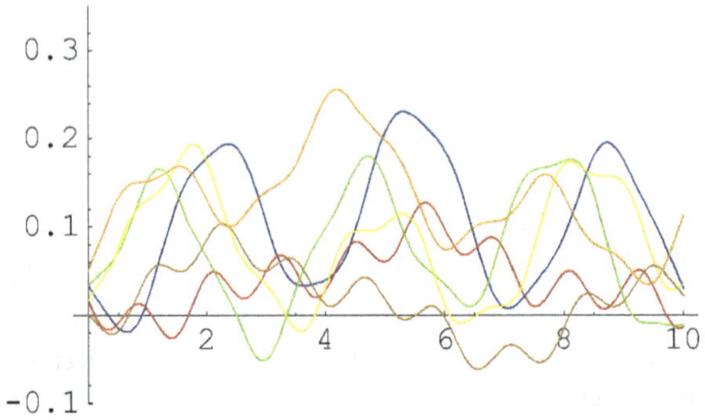

Fig. 6.63 Player strategy–flows (vertical axis) versus path length (horizontal axis) for a three person game—Player one strategy–flows are blue and red, player two strategy–flows are green and yellow, player three strategy–flows are brown and orange and time is off scale.

The flow reflects this same structure, Fig. 6.63.

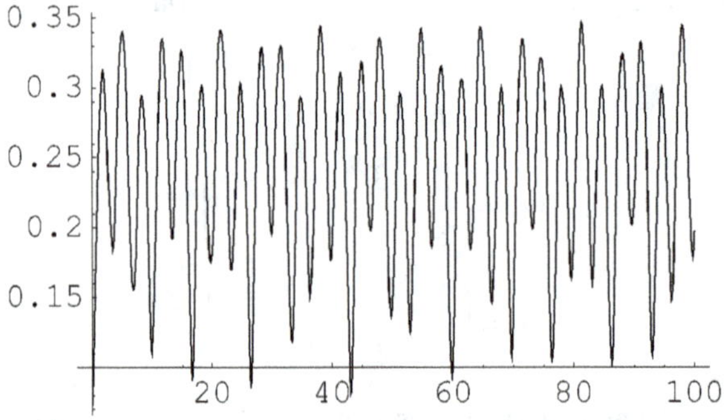

Fig. 6.64 Market boost (vertical axis) versus path length (horizontal axis) presented for a three person game.

The energy of the system is conserved and so the market boost Fig. 6.64 is bounded.

The market boost oscillates, reflecting the existence of opposing forces. Equation (6.5) shows that market speed plays off the three-person ***game bias*** A_0. Although the magnitudes of the coalition biases increase, the game bias is bounded since the market boost is bounded.

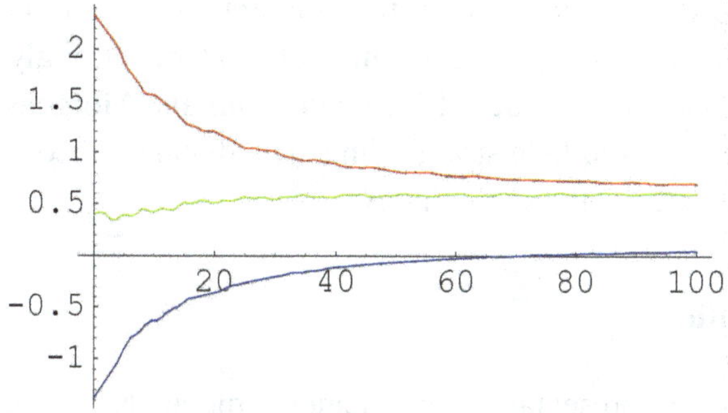

Fig. 6.65 Imputations (vertical axis) versus path length (horizontal axis) for a three person game—blue, red and green represent player one, two and three respectively.

The payout to each player is set by the initial conditions, which I have taken to be a defensive strategy for each player.

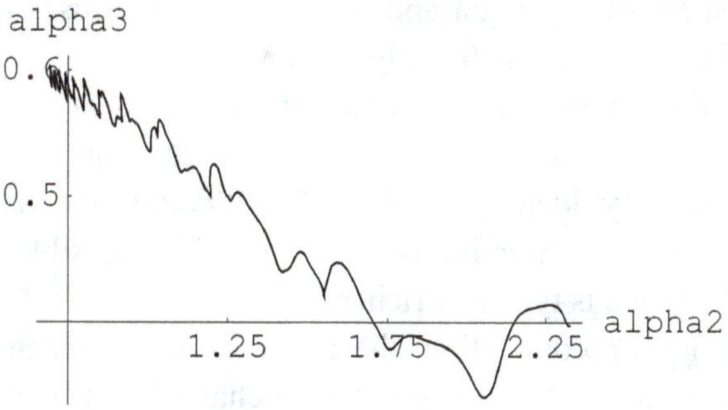

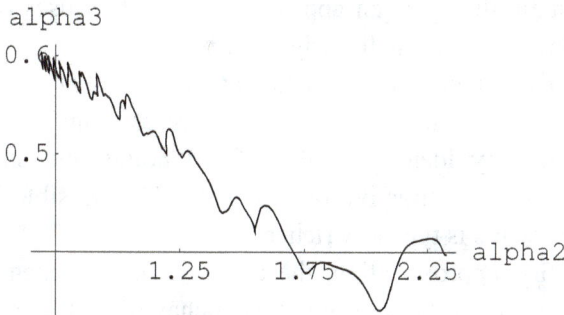

Fig. 6.66 Player-three imputation (vertical axis) versus player-two imputation (horizontal axis) presented for a three person game.

The payouts (imputations) are shown in Fig. 6.65.

In principle the payout to each player will equal or exceed the defensive amount by an amount called the player's ***imputation***, the difference between the value for this player and the defensive value of the game where this player is pitted against a coalition of the other two players.

It appears that player one gets left out, receiving a little more than his defensive position. The sum of the imputations is constant (for this example, the sum is $4/3$). The imputations are strongly correlated, Fig. 6.66. In the static model of Von Neumann and Morgenstern (1944), this correlation would be a straight line. The dynamic theory for a three-person game extends the static theory.

6.4 Observations

The graphical presentation complements the analytic discussion from the previous chapter. The two discussions make clear certain general attributes of the dynamical game theory. I summarize these attributes with the following observations:

1. The static game theory expectations of Von Neumann and Morgenstern (1944) appear in the dynamic theory as preferred directions in space–time corresponding to null vectors of the market field.
2. The static theory is an approximation that focuses on strategies that increase almost linearly with path length. Such behavior is indicative of inactive strategies and/or stationary flows that are fixed points. In general, one analyzes complex differential equations by identifying their fixed points and stable curves, which can be attractive or repulsive. The possible behavior of such systems is typically richer.
3. To a first approximation, the dynamic theory extends the static theory with a basic oscillatory behavior. At a deeper level, angular isometries will be reflected in an oscillatory behavior that is in general not periodic. Such ***chaotic*** behavior is reflective of the highly non-linear nature of the equations.
4. In the dynamic theory the value of the game has strategic consequences in contrast to the static theory.

5. Another significant departure from the static theory is the presence of gravitational, pressure gradient and market bias forces. These result from treating dynamic games as charged fluids in motion. Fluid parameters characterize realistic attributes of games that are more psychological than economic. For example I associate "experience" or learning with the strategic-density; "control" with pressure. These attributes can play as significant a role in the final behavior as the pure market forces.
6. A general result for theories with a time isometry is the conservation of "energy" or "time–charge". The approximate form is Eq. (6.5), the generalization to strategy–space–time of the Bernoulli equation in classical fluids that constrains the system response to motion or market boost, gravity, pressure and market bias.
7. The speed of the strategic-density flow, the market boost, plays an essential role. It is also worth noting that only the market bias of the game matrix enters in the "energy" conservation. The player's payoffs do not.
8. The game matrix or payoff matrix determines the general oscillatory behavior. I interpret this oscillatory behavior as normal player uncertainty about where to place the strategy–choices. In the sense of a table at roulette, the players bet and can not only add chits to a square representing each strategy but can take away chits. At any point in time, a player may show more confidence about one strategy versus another. Such behaviors reflect real world behaviors.
9. The oscillatory structure reflects distinct periods. In the models studied, these periods are reflective of the properties of the original game matrix. It reflects the values of the payoff matrix, the number of eigenvalues, as well as the interplay between the varying forces of market, pressure and gravity. When the "angular momenta" are conserved, the structure also reflects the additional angular momentum conservation laws.
10. Gravity initially moves the system toward the static equilibrium point. The system is set into an oscillatory motion as a consequence.

11. Pressure initially moves the system away from the equilibrium position, not toward it. The system is set into an oscillatory motion as a consequence. Pressure is essential since it keeps the system from collapsing. This is consistent with pressure being a short-range force.
12. The parameters of the theory such as pressure and gravity are functions of the active strategies. Assumptions about the functional form of these parameters provide distinctions about the behavior of economic systems. The single strategy model gives initial insight into these functional forms. This suggests that for two or more active strategies, numerical solutions of the full equations be studied.
13. The periodic or semi-periodic structure extends the notion of stable behavior introduced by Von Neumann. The system is not moved to the equilibrium position and then left there; the system moves dynamically in response to forces. Dynamic systems of the type discussed here will have in general both stable and unstable fixed points.
14. In particular, stable behavior for three or more persons differs substantially from the static theory. For such games, the defensive position is not a stable point in general. The theory suggests how coalitions interact to produce stable forms of behavior. These forms are in general not fixed points but chaotic orbits.
15. The concept of coalitions in such games needs further study. The initial indications are that imputations are similar in concept to the static theory but different in execution.
16. The restriction to positive strategies is required if normalized weights are used to computed what are normalized values. This could be relaxed by introducing un-normalized values computed by un-normalized strategies. Such un-normalized strategies can be positive or negative. This might be more in keeping with the idea that strategies can increase or decrease.
17. Game value was "measured" in a non-covariant way using strategies in order to connect with the usual notions of value. A better measure of game value might be a covariant definition.

Chapter 7

Applications and Open Problems

There are many possible problems to explore at this point. My ultimate goal is to construct credible models and apply them to realistic problems. Based on the scientific method, this will force the theory to adapt to experimental input. I see the need to expand the theoretical discussion of games for three or more players. And of course, I think it essential to continue to strengthen the theoretical foundations. With these goals in mind, I list applications and open problems that I find interesting that could be the subject of future investigations. I am sure there are additional questions that may occur to the reader, so my list is not exhaustive.

7.1 Organizational Dynamics

I have spent more than 20 years in the software industry, in both small and large companies. These companies produced software products (such as lines of code, documentation, *etc.*) using lots of people and lots of managers. I was struck by the simple rules that seemed to govern the project management of these organizations [See e.g. DeMarco (1982) and Brooks (1982)]:

- Each manager has a ***span of control*** N that is roughly constant across the organization.
- Organizations are hierarchical so that the size of the company is determined by the managerial overhead or number of ***layers*** k.
- The ***workers*** appear at the bottom of the hierarchy, so the number of workers $W = N^k$ is determined by the span of control and the

number of layers. Everyone else in the organization is a ***manager***. Managers produce no output.

- At each level of the organization, the ***output*** is determined by the span of control times the ***unit-productivity*** or ***effectiveness*** E : $O = N \times E$. This is the ***production hypothesis***.
- The ***total output*** $O_{\text{total}} = O^k$ in terms of such things as lines of code per year, pages of documentation per year, *etc.* of the organization is proportional to the number of workers.
- The ***mythical man–month hypothesis*** holds: $W = O_{\text{total}}{}^{1+\varepsilon}$. Organizations are characterized by this positive ***overhead*** ε: Doubling the size of the job requires more than just double the number of people.
- The mythical man–month can be expressed in terms of the span of control and the output: $N = O^{1+\varepsilon}$.

One job of a manager is to provide control and manage resources, including people, to deliver a given job with a given size on a specified date. This is one possible definition of project management. The attributes above are the essential ingredients to accomplishing these goals.

I see the possibility that these attributes might be explained in terms of the dynamic theory. I think of an organization as multiple two-person games between a manager and his/her direct reports at each level of the hierarchy. There are opposing goals between workers and managers and the opposition reflects itself in the attributes of the organization described above. I believe the organization can be characterized by its strategic-mass ρ, market control, p and its thermal properties as measured by the temperature (measuring the "internal" degrees of freedom) T and that these quantities are some permutation of the output, span of control and effectiveness.

In this simple exposition there are twelve possible ways to identify pressure, density and temperature to span of control, effectiveness and output. Each of these ways will convert the ***production hypothesis*** to the perfect fluid law $p = \rho T$ and the mythical ***man–month hypothesis*** into the adiabatic flow relationship $p = \rho^{1+\alpha^{-1}}$. The dynamic theory provides a possible origin for the production and mythical man–month hypotheses.

The market overhead index α will be determined by the overhead ε, which is positive. Of the twelve identifications, there are only four identifications that satisfy the requirement that the market overhead index be positive always:

p	ρ	T	α
N	E^{-1}	O	ε
N^{-1}	E	O^{-1}	ε
N	O	E^{-1}	ε^{-1}
N^{-1}	O^{-1}	E	ε^{-1}

I make the following provisional identification: The market overhead index is equal to the overhead. I note that the overhead α in a perfect fluid determines the specific heat at constant volume (constant strategic density): It determines how much heat a system will gain for a unit change in temperature (resource). A system with a large market overhead index gains a great deal of heat from a unit change in temperature. In other words it has more internal degrees of freedom. A high market overhead is then associated with systems which can obtain lots of heat quickly for a given temperature change. Since I associate heat with energy that does no work, it means I associate such systems with high overhead.

With this identification, the two choices are then the first two in the table. These are reasonable from the perspective that the strategic-density can be reasonably associated with effectiveness or unit-productivity. I believe it makes sense to associate more effectiveness with more strategic-density. The choice is then the second one and captures my own direct observation that increased control is inversely proportional to the span of control:

- Control (pressure) is inversely proportional to the span of control, $p = N^{-1}$;
- Strategic-density is equal to effectiveness or unit-productivity, $\rho = E$;
- temperature is the ***resources*** needed to output one unit, $T = O^{-1}$;
- And the market overhead index is equal to the overhead, $\varepsilon = \alpha$.

The single strategy model suggests that at the market equilibrium there is a maximum control (pressure) and therefore a minimum span of control. This is consistent with local wisdom. Organizations in trouble want to increase their effectiveness or unit-productivity (their strategic-density) and do this through increased management control by lowering the span of control at constant output. The single strategy model shows that in thermal equilibrium larger spans of control correspond to larger outputs (lower temperatures) and lower effectiveness or unit-productivity. According to the single strategy model the temperature distribution would be characterized by constant $T\sqrt{\gamma_{00}}$.

Thus a dynamic theory suggests that organizations in general will not be in economic static equilibrium but in thermal equilibrium. If so they would be ***stratified***. Upper management would be close to the optimal game equilibrium and be characterized by a high strategic-density (effectiveness), high temperature (number of resources to output one unit or "managers produce nothing") and high pressure (low span of control). Workers however would be characterized by low strategic-density (effectiveness), low temperature (number of resources to output one unit or "workers produce everything") and low pressure (high span of control). Such an organization would align its component parts (groups, divisions, business units, *etc.*) on a curve whose components obey the mythical man–month relation, all characterized by the same market overhead index.

An organization with low market overhead index behaves like a fluid with few degrees of freedom. There will always be some overhead in an organization. Some effort goes into communicating to workers, educating, providing benefits, counseling, *etc.* I have observed market overhead index values in the approximate range of $\frac{1}{10}$ to 2.

7.2 Reorganization Cycles

Business goes through ups and downs and so organizations respond with their corresponding ups and downs. I have always been curious why organizations in particular respond to such cycles with their own downsizing and upsizing cycles and I think the proposed dynamic theory can address this question. I think the hint is that reorganizations operate as a thermodynamic heat engine. There are four distinct cycles in the $W - O_{\text{total}}$ plane, using the notation above:

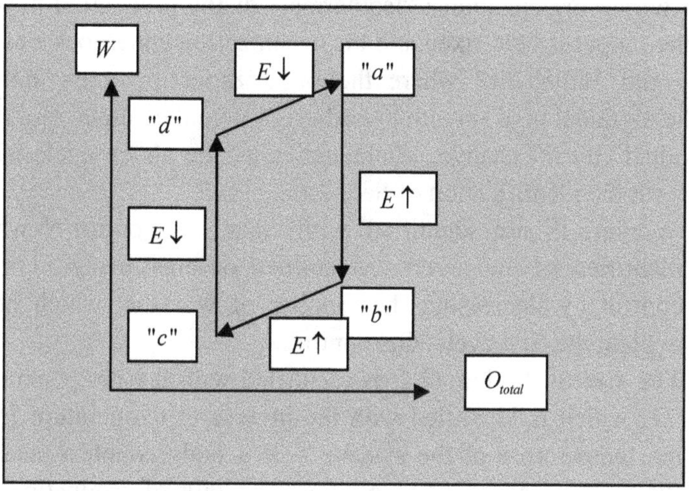

a) At constant output ("isothermal"), the organization decreases the number of workers improving efficiency or unit-productivity. This happens outside of normal business practice, often instigated by the Board of Directors.
b) The Board now allows the market forces to take over and the organizational dynamics proceeds as normal and so the mythical man-month hypothesis applies ("adiabatic"): The output drops along with a continuing drop of the number of workers. The size of the organization is less, though with a still increasing unit-productivity. The organization moves towards market equilibrium.
c) The organization is now operating optimally and because of success, the work force grows. Such growth is outside of the market forces because the output is held fixed ("isothermal"): The unit-productivity decreases.
d) Business practices take over and the output is increased, with a corresponding and continuing increase of workers and decreasing unit-productivity ("adiabatic"). At some point the organization is

back where it started from, understanding it is far from economic equilibrium and in trouble.

The above cycle depends on two behaviors generally understood to operate within organizations: "Isothermal" or the production hypothesis where the output is held fixed and the organization undergoes a profound change; and "adiabatic" where the organization operates "normally" under the mythical man–month hypothesis where the output, workers and unit-productivity all change. Mathematically, the above cycle is a heat engine with the identification of Sec. 7.1:

a) Workers W are identified with Span of Control N which is identified as the inverse of ***control*** or pressure p. I increase control by decreasing the number of workers, which helps to explain the first cycle: Get control.
b) The size of the job O_{total} is identified with the per person output O, which is identified with the inverse of temperature T. Thus the temperature of the system T is a body count, a measure of the ***resources*** needed to produce one unit of production. In the first cycle I keep the resource measure fixed (isothermal) while focusing on increasing control.
c) The unit-productivity or effectiveness E at each level is identified with the political ***capital*** or strategic-density ρ. The goal in the first cycle is focused on the fact that the constant output is equal to the unit-productivity times the number of workers, so by reducing the number of workers a higher unit-productivity is obtained.
d) Under normal business conditions, the unit-productivity, the output and the number of workers all change subject to the mythical man-month relationship: $W \propto O_{\text{total}}{}^{1+\varepsilon}$. By identifying the overhead ε with α, I identify the mythical man–month relationship with the adiabatic relationship between pressure and temperature: $p \propto T^{1+\alpha}$. The usual meaning of adiabatic change is that the entropy of the system does not change. The second and fourth cycles are ones where typical business practices operate and so the system changes are adiabatic.

I note that the efficiency of a heat engine is equal to the temperature difference divided by the maximum temperature. Thus the efficiency of the organizational heat engine is equal to the difference in the maximum and minimum job size divided by the maximum job size. Moreover the heat engine does work. I propose that the work done for an organization is to generate flow of strategic-mass, *i.e.* a "wind" in choice–space; the greater the amount of work, the greater the wind and its possible side effects. One possible side effect is the increase in strategic-density for some strata. I think of the analogy with weather. The heat engine initiated by the sun generates weather patterns on earth which indeed produce significant stratification in the form of clouds and rain.

7.3 What is a Player?

In this monograph I distinguish a dynamic game theory as a theory of choices made by players. I have endowed players with a value–choice, whose direct consequence is an associated charge and market field distinct from any other player. In the single strategy model, I computed each player 's self-interest that results (*Cf.* Appendix E) and chose initial conditions so that this was positive. This is the normal sense that each player would optimize their ***self-interest***. These players might be called ***sharps***. There is a logical possibility however for players that minimize their self-interest, who might be called ***flats*** or ***altruists***. The last logical possibility is that there may be players that have no charge, which might be called ***naturals*** or ***dropouts***.

These possibilities are important since different combinations of these sharps, flats and naturals will produce qualitatively different games. These distinctions must be part of any definition of a player. In this monograph, I have assumed that the players are all sharps and that they see the same game. This needs not be the case, even if they are all sharps. One player might "see" a game with much greater strength than the other players and so the effective game will be that of the stronger player. With the possibility of flats and naturals, there are even more possibilities.

In addition to these distinctions, I have made the assumption in the monograph that there were no factions; I assumed the usual game form in which the idea of the player is reflected in the form of the market field. I now inquire whether there is some principle that enforces this. In particular I inquire whether there is a specific form of the sources (energy-momentum tensor) that dynamically distinguishes a player.

I start by supposing there is a transformation group decomposition[42] of the space into N players:

$$R_0 \oplus R_1 \oplus \cdots \oplus R_N \, .$$

This might provide an operational definition of a ***player***. For example, the transformation group for each subspace might be the product of a one-dimensional inactive value–choice times a projective group: Each player would have a number of strategies that could be considered homogenous coordinates of this projective space.

In this formulation, for each player I would relate each strategy–choice x^a to the new variable $y^a = \ln x^a$. The metric depends on the difference of the new variables. Furthermore, I would expect the metric to be block–diagonal, as well as the market potentials that result from the market potentials $A_a^{\bar{k}}$ for each player. This is the formulation made in this monograph; here I interpret the player by the transformation decomposition into a product of projective spaces.

Since I draw the distinction that strategies are associated with a given player, I am led to the problem of making that same distinction when writing the energy–momentum tensor source from the metric. To get a sense of what to expect, I note that the player decomposition would be consistent with the following metric form:

$$g_{\mu\nu} = \begin{pmatrix} g_{00} & g_{0I} & g_{0II} \\ g_{0I} & g_{I,I} & 0 \\ g_{0II} & 0 & g_{II,II} \end{pmatrix} .$$

This is a ***player block diagonal*** metric. To make this into a covariant expression, I must relate the statement about the metric to a statement about the curvature and then ultimately to the sources. The curvature will depend not only on the metric form but on the form of the market field contributions.

---

[42] The first term is for time, $t = \xi^0$.

To begin however, I might hope that the full curvature tensor (including both active and inactive market contributions) would break into the same type of decomposition:

$$R_{\mu\nu} = \begin{pmatrix} R_{00} & R_{0I} & R_{0II} \\ R_{0I} & R_{I,I} & 0 \\ R_{0II} & 0 & R_{II,II} \end{pmatrix}.$$

It would then follow that the energy momentum tensor has the same decomposition:

$$T_{\mu\nu} = \begin{pmatrix} T_{00} & T_{0I} & T_{0II} \\ T_{0I} & T_{I,I} & 0 \\ T_{0II} & 0 & T_{II,II} \end{pmatrix}.$$

A little thought however shows that since the metric depends on the choices of all the players, it will in general not be true that the curvature tensor has the block diagonal form.

The value of looking at the sources however provides an alternate approach. I start with the assumption that each player can apply stress only to their choices in the central frame. I assume at the outset that the energy momentum tensor has the above block–diagonal form with a distinct pressure scalar for each player. This is described in Appendix G as the ***Player Independence Hypothesis***. The metric and market field then result from a solution to the full set of field equations.

The specific form is called a ***player fluid*** for which I have found no analogy in physics. I think a study of such fluids and a further study of the types of players (sharps, flats and naturals) might provide insight into possible economic behaviors.

7.4 Flat Games

The specific distinction of two-person games is the possibility that the equilibrium point is not only a fixed point but a stable point. Two-person games are special cases of flat games for three or more persons. A ***flat game*** is one in which there is no advantage to form a coalition. I consider what happens at the defensive position for such games.

To illustrate the idea, consider a game with any number of players. I consider each player competing against all other players in coalition. Under these circumstances, I define a strategy $\boldsymbol{\eta}^\alpha$ for that player. This is a vector in the sub-geometry, which I normalize so that the sum of the components is unity. I take the external sum of these vectors, along with a time component, to form a flow vector $\eta^a = \oplus\sum \boldsymbol{\eta}^\alpha \oplus \{m_0\}$. If player α plays his optimum strategy, then no matter what the coalition does, he receives at least the ***defensive payoff*** v_α:

$$\eta^{\alpha(n)} f_{nl} \geq v_\alpha .$$

Now consider any set of $N-1$ players, say $\{2,\cdots,N\}$. They will receive potentially more, $-v_1$, if they join in a coalition than they would receive if they acted alone, namely $v_2 + \cdots + v_N$; in other words $-v_1 \geq v_2 + \cdots + v_N$. So in particular, the defensive playoffs satisfy the rule:

$$\Delta \equiv -\sum_\alpha v_\alpha \geq 0 . \tag{7.1}$$

A game is ***flat*** if the deviation above is zero. A two-person game is always flat and there is reason to believe that the defensive point is not only fixed but stable: Nearby behaviors are attracted to this point. For games that are not flat, there is no reason to believe this. However, for games with three or more persons, a flat game might in fact have a stable fixed point. The structure of such games needs to be further clarified.

7.5 Three-Person and Higher Games

An important development of the theory will be to explore games with more than three players. In such games I have demonstrated that coalitions play a central role though in a way that is different from the static theory. Thus this dynamic theory might give useful insight into the nature of games with multiple players. A preliminary study of such models suggests that games with multiple players might be amenable to techniques from physics for dealing with many-body problems. The potential for such systems is constructed from a sum of all the two-body forces. Here the "two-body forces" can be thought to be determined by the coalitions. For game theory, it may be possible to consider the market potential as being made up of the sum over all the possible coalitions.

I have articulated a theory for any number of players. I have explored very little of these games. It may well be that on examination, there will be issues prompting a refinement of the theory. This was certainly the case for the static games considered by Von Neumann and Morgenstern (1944).

7.6 Non-Zero Sum Games

There are extensions to the theory that have profound impact in understanding real markets. I have formulated the theory for both zero-sum and non-zero sum games. As with the static theory, a non-zero sum game can be thought of as a zero sum game with one more player. I think it would be fruitful to consider such games in more detail. The theory might provide insight into games such as the Prisoner's Dilemma [*Cf.* Luce and Raiffa (1957)]. I conjecture that a more careful analysis might not require the introduction of a fictitious player.

7.7 Viscous Games

It is possible to consider a different generalization to the perfect fluid than the one in Sec. 7.3. I think of the market fluid as the attribute of a market defined as a collection of games being played. The games are played multiple times and so it makes sense to consider the set of games as defining a flow; the flow consists of successive points of play for each game in the market. Although the rules-of-the-game determine the play, I posit that the players of the games do communicate outside the rules-of-the-game. These communications are complex and can be presumed to be described by complex and competing positive and negative feedback loops. I propose to capture the net effect of these complexities by ascribing ***substance*** and ***elasticity*** to the games. The substance responds to "body forces" and the elasticity represents the internal forces between games in the market. The substance and elasticity model the communications of the players that lie outside the rules-of-the-game.

I have described a dynamic theory of games with body forces and simple elasticity. The "body forces" subject the substance to motion that is independent of the market. The market forces are a result of the substance having a charge and reflect the rules-of-the-game. With simple elasticity, I assure that the system does not collapse to a point. With more complicated models of elasticity, the games also flow with "internal" resistance. It is the internal resistance that will generate substantially new phenomena.

What is the origin of this "internal" resistance? My idea is that the complex positive and negative feedback loops that characterize the communication between players which lie outside the rules-of-the-game can be described by systems dynamics concepts. Systems dynamics models these communications by hordes of coupled differential equations and as such could itself be modeled as a bunch of interacting springs with various possible connections. But mathematically such a bunch of interacting springs is also a pretty good model (at least a thought model) of a classical solid or fluid as understood a couple of hundred years ago before the prevalence of the atomic theory.

I come to a theory then that approximates the real world. It is a continuum model that has both substance and elasticity. In such a model, I would expect even an isotropic fluid to have in addition to a ***market pressure***, a ***market viscosity*** that models the internal forces. Moreover, a realistic model would have thermodynamic attributes such as temperature, heat and thermal conductivity [*Cf.* Appendix A]. Given our knowledge of real fluids, it is clear that these new attributes of the fluid will provide phenomena that are substantially different from an ideal fluid.

7.8 Quantum Games

There are other aspects of the general theory that need exploring. As an example, I have suggested a theory not unlike electromagnetic theory in physics. The analogies suggest additional development. As an example I have described cyclic behavior based on the forces. I have ascribed a "charge" to a game and so the orbit consists of a charged particle going around a circle. Such behaviors in physics are not possible without the charged particle radiating energy. That should be true in this theory as well. Under these circumstances it may be that the game will indeed radiate. That suggests an inquiry into the nature of the radiation. It may also be that not unlike atoms, charged particles get trapped into orbits but don't radiate. This would lead to a development in the theory analogous to quantum mechanics. The theory could be reformulated using techniques from these other areas. Modern arguments about quantum mechanics would modify the essentially classical theory described here.

7.9 Complete Solutions

I have one example where all of the field equations are solved; the single strategy model. I believe it is essential to consider solutions to the full equations to gain a proper understanding of the theory. One such approach might be to use numerical techniques. If possible, I believe this would be the most fruitful.

However, there might be other approaches that give approximate solutions to the full field equations without assuming all strategies are inactive but one. For example, I observe in Appendix C that to obtain a solution to the field equations, a specific gauge needs to be specified. I also indicate that there is a weak field approximation to the field equations that can provide insight, though I note that in the single strategy exact solutions, gravity effects are large in economics.

I provided in the monograph detailed analyses of two-person games using the equations governing the flow; the "Newtonian" equations of games. One approach to the flow equations is the study of streamlines. They provide a coupled set of equations for the flow and its gradients and hence can lead to a solution to the problem of computing the flow assuming the metric is known. In the central frame, I believe the sources for the behavior will be the scalar gravitational gradients and the pressure gradients; components of the pressure gradient are determined by the deviation equations and the conservation of strategic-mass requirement. Given the machinery provided, it is possible to take a given game, provide a suitable metric that defines the rational and market forces and study the resultant flow. The flow will provide the distinctions between each player and each player's strategy.

Nevertheless, such approximations may overlook interesting effects. Full solutions to the field equations for dynamic games for example might open up the possibility of observing new and interesting phenomena. For example, I have considered only stationary metrics, *i.e.* the metric independent of time. That condition should be relaxed in realistic game situations. When that happens, one might see that "gravity" or "markets" can propagate as waves. This follows from Eq. (C.13). Such phenomena would have interesting interpretations in economics.

Appendix A

Thermodynamics

In a sequence of plays or between neighboring plays, I assert that there are elastic forces. I further suggest that an initial step in understanding such forces is to describe the medium as a fluid. Great progress was made understanding fluids using thermodynamics. Will such an approach prove useful here? I believe it will. In the literature, there exists a treatment of thermodynamics that focus on its geometrical aspects. The treatment is by Tolman (1987) and lays out the assumptions needed for a thermodynamic theory in such a way that it can be embedded in an arbitrary geometry. I generate the argument here in the frame work of such geometry, with comments appropriate to the theory of games.

Tolman introduced thermodynamics into the relativistic theory as a geometric distinction. In that theory there is temperature, entropy and heat. The Oxford English Dictionary (2002) describes thermal as "Of, pertaining to, or of the nature of *thermæ* or hot springs; of a spring, *etc.* (naturally) hot or warm; also, having hot springs." Thus thermal denotes heat. Thermal equilibrium means that heat does not flow from one part of the system to another. Classical physics distinguishes two types of energy: mechanical (*i.e.* energy associated with motion) and heat (*i.e.* energy associated with internal degrees of freedom not specifically evidenced by motion).

In a geometric theory, the first law of thermodynamics is captured by the conservation of energy: In our relativistic theory of games it is the statement that the energy momentum tensor is given in terms of the curvature, which is determined by the sources, Eq. (1.2). The distinction between the two types of energy is not visible: Whatever heat is, its effect can only be to influence the sources and the metric and it must do so in such a way that the sources determine the metric. Tolman provides the analysis of the distinction between these two types of energy.

He starts with the principle of the first law, which is to express"...the principle of the conservation of energy by equating the total energy change in a system to that which is transferred across the boundary; and is to be regarded in the second place as introducing a distinction between the two methods of energy transfer—***flow of heat*** and ***performance of work***—which becomes especially important for the later application of the second law of thermodynamics."

In order to distinguish between the two methods of energy transfer, Tolman proposes a new scalar field, the ***entropy density***: The second law of thermodynamics provides the distinction of ***entropy*** as a relationship between the entropy density, flow and volume element on the one hand and the ratio of the heat and temperature on the other hand:

$$\left(\varphi \frac{dx^\mu}{ds}\right)_{;\mu} \sqrt{|g|}\, d^D x \geq \frac{\delta Q}{T}. \tag{A.1}$$

What is new for the theory of games is the introduction of the general volume element instead of the four dimensional space–time of physics and the meaning given to the strategic-space–time dimensions of the geometry. The resultant notion of temperature is well defined for our dynamic theory of games.

Equation (A.1) introduces the heat transferred across the boundary and the temperature on the boundary. This expression is related to the metric and sources. Gauss' theorem can be applied to the second law:

$$\int_{\partial V} \varphi \frac{dx^\mu}{ds} d\sigma_\mu \geq \int_V \frac{\delta Q}{T}.$$

This says that the entropy enclosed in a space-like region changes along the integral curve to the flow if heat is added to the system. The change exceeds the heat added if the process is irreversible. Heat is associated with the flow of entropy, which is the new geometric structure.

I make the metric dependence more explicit. The time-like flow vector defines an integral curve through any given point. The entropy, energy and volume of a thin shell of the system at that point are determined:

$$S = \varphi \frac{dx^\mu}{ds} d\sigma_\mu$$
$$E = \mu \frac{dx^\mu}{ds} d\sigma_\mu \, . \tag{A.2}$$
$$\mathbf{V} = \frac{dx^\mu}{ds} d\sigma_\mu$$

The energy density defined here is given by the sources Eq. (1.4):

$$\mu = T_{\mu\nu} \frac{dx^\mu}{ds}\frac{dx^\nu}{ds} \, .$$

I take this to be true even if the form of the energy momentum tensor is not given by the perfect fluid. The pressure is likewise defined in terms of the energy momentum tensor, even if the fluid is not a perfect fluid:

$$p = \frac{1}{D-1} T_{\mu\nu} \left(\hat{\gamma}^{\mu\nu} - \frac{dx^\mu}{ds}\frac{dx^\nu}{ds} \right) .$$

These quantities can be computed from the curvature tensor by means of the field equations. They allow a definition to be given for changes to the total energy due to changes in the metric and explicitly allow the distinction to be made between heat and mechanical energy:

$$\delta E = T\delta S - p\delta \mathbf{V} \, .$$

Mechanical energy is determined by the pressure and volume changes which necessarily result from the motion of the fluid; total energy change is given by the change to energy density times the volume; heat $T\delta S$ is what is left over.

Let us consider changes in the system generated by changes in the metric along the integral curve defined by the flow. These changes in the system do not affect the flow and are orthogonal to the flow. For a system in thermal equilibrium, the change in the entropy along the path is related to the energy and pressure of the system for a differential slice:

$$\delta S = \frac{1}{T}\delta E + \frac{p}{T}\delta \mathbf{V} \, .$$

The changes are summed (integrated) over a finite space-like part of the system:

$$\begin{aligned}&\delta\int S = \int\frac{1}{T}\delta(\mu\mathbf{V}) + \int\frac{p}{T}\delta\mathbf{V} = 0\\&\therefore \int\left(\frac{\mathbf{V}}{T}\delta\mu + \frac{\mu+p}{T}\delta\mathbf{V}\right) = 0\end{aligned} \qquad \text{(A.3)}$$

I now consider a specific region of interest and consider the restrictions imposed on that region assuming there are no changes on the boundary of the region and in the system outside the region of interest. This means there are no changes in the energy momentum at the boundary or outside the region and no changes in the sources. In other words, the first and second derivatives of the metric are unchanged on the boundary and outside the region of interest:

$$\delta\gamma_{\mu\nu} = \delta\left(\partial_\lambda\gamma_{\mu\nu}\right) = \delta\left(\partial_\delta\partial_\lambda\gamma_{\mu\nu}\right) = 0\,.$$

I imagine that along the path, the system orthogonal to the path is unchanged up to a specific point. I allow changes to occur for a small distance after that point. After that small distance, I allow no changes to the system. The changes are made only by changing the metric subject to the conditions on the boundary and outside the region of interest.

The condition for ***thermal equilibrium*** is that the space-like integral of the entropy density not change along the integral curve, since the assumption is that once the integral is performed over the orthogonal space-like region, there is no heat transfer across the boundary as a whole:

$$\delta\int\varphi\frac{dx^\mu}{ds}d\sigma_\mu = 0\,.$$

However, within the region of interest, changes can occur that involve either heat transfer or mechanical work. Whatever that mixture, the total energy is conserved. The ***heat transfer*** is measured by $T\delta S$ and the ***mechanical work*** by $-p\delta V$. The variations are induced by changes in the metric and its derivatives subject to the conditions on the boundary and outside the system.

The energy density and pressure are constrained by the metric and so Eq. (A.3) provides a fairly complex set of conditions or constraints. The temperature is an unknown function, an integrating factor, which makes a solution possible. I have not solved these equations in general but have found solutions for the single strategy model [*Cf.* Appendix D]. Tolman (1987) suggests that under certain assumptions, $T\sqrt{\gamma_{00}} = const$, which I put in the form:

$$\frac{1}{T}\frac{dT}{dp} = \frac{1}{\mu + p}. \tag{A.4}$$

The important conclusion to be drawn is that the ideas of thermodynamics can be borrowed and applied to the theory of games. Though I haven't proved it, the market is defined by two functions, pressure and energy density. To the extent thermal equilibrium leads to an equation such as Eq. (A.4), a third function, temperature, can be introduced. The elastic fluid Eq. (1.3) is described in the literature as an ***isentropic*** fluid: This is a fluid in which the entropy does not change. Indeed for all solutions of the equations of motion, the entropy does not change along a streamline path.

As an illustrative example, assume that the energy density is proportional to the pressure:

$$\mu = \alpha p. \tag{A.5}$$

The temperature can be computed from Eq. (A.4):

$$\ln T = \frac{1}{1+\alpha}\ln p + const. \Rightarrow p = const.\times T^{1+\alpha}.$$

Equation (1.3) can be slightly rewritten.

$$\frac{1}{\rho}\frac{d\rho}{d\mu} = \frac{1}{\mu + p}. \tag{A.6}$$

From this equation, I get the density as a function of pressure:

$$\ln \rho = \frac{\alpha}{1+\alpha}\ln p + const. \Rightarrow p = const.\times \rho^{1+1/\alpha}.$$

This equation is normally associated with an ***adiabatic*** fluid, namely a fluid where the expansion occurs without any change in heat. There will be a change in temperature. Such changes are important to heat engines and hence such processes as internal combustion engines.

These equations provide a simple relationship between pressure, density and temperature:

$$p = const. \times \rho T .$$

This is the "equation of state" for a perfect fluid. It is a relation between pressure and density (or for pressure and temperature) for fluids undergoing adiabatic changes (no net entropy changes). The perfect gas law does not hold in general when the energy density is a more general function of pressure.

In the example above, the index α plays an important role and thus provides an important distinction for games. I define it more generally as $\alpha \equiv d\mu/dp$ if I let energy density be a general function of pressure. The larger the index, the more the system compresses:

$$\frac{p}{\rho}\left(\frac{\partial \rho}{\partial p}\right)_S = \frac{\alpha p}{\mu + p} . \tag{A.7}$$

I call the index α the ***market overhead index***; it determines the compressibility of the fluid when no heat flow occurs.

By focusing on the sources, I am able to borrow a good deal from the mathematical literature on thermodynamics concerning elastic fluids. The caveat is that the borrowings are from the behavior of relativistic fluids, such as one might study in astronomy, not ordinary fluids. The reason is that heat and mechanical work are both forms of energy and both are sources for the metric. The metric in relativistic theories is associated with gravity. Thus heat can generate gravity. This modifies the common notion that a system in equilibrium has a constant temperature. In fact, in simple models $T\sqrt{\gamma_{00}}$ is constant. The time component of the metric is directly related to gravity and becomes small when gravity becomes large. This says that in areas where gravity is larger, temperature is larger. For games, we will see that the game equilibrium is expected where the "gravity" is large and so the temperature will be correspondingly larger there than away from game equilibrium.

To extend the discussion from models of elastic fluids to more realistic models in which there are both viscosity and heat flow [see for example Chandrasekhar (1961)], the literature suggests including in the energy momentum tensor terms that describe both:

$$T_{\mu\nu} = (\mu + p)V_\mu V_\nu - p g_{\mu\nu} + \eta\sigma_{\mu\nu} + k\left(h_\mu^\lambda V_\nu + h_\nu^\lambda V_\mu\right)\partial_\lambda \varphi . \quad \text{(A.8)}$$

The idea is to provide an expression that conforms to general experience in a frame that co-moves with the flow. For this reason there is a projection operator $h_\mu^\nu = \delta_\mu^\nu - V_\mu V^\nu$ for the temperature gradient. The convention chosen for the energy density is $V^\mu V^\nu T_{\mu\nu} = \mu$. The convention chosen for the pressure is $h^{\mu\nu} T_{\mu\nu} = -(D-1)p$ where the number of dimensions of space–time is D. The above form of the energy momentum tensor is defined with the first two terms representing the "perfect" fluid, the next term representing viscosity as resulting from a contribution proportional to the shear strain (shear has $h^{\mu\nu}\sigma_{\mu\nu} = 0$; see Appendix F) and the last term representing the contribution coming from heat flowing within the fluid with a contribution proportional to the gradient of a scalar φ that measures the heat flow.

In non-relativistic physics, the scalar φ is the temperature. For a system in equilibrium, the system has a constant temperature. Heat flow is characterized by changes in temperature. In the models considered here however, a system in thermal equilibrium need not have a constant temperature [*Cf.* Appendix D]. No heat flow is characterized by a lack of change in entropy. In the single strategy model example, there are also solutions with entropy change that are in thermal equilibrium. Hence heat flow is taken into account automatically using the formalism from this appendix. I therefore take $\varphi = 0$.

The internal energy is no longer assumed to be simply related to the pressure and market density. Market density or strategic-mass can be defined in terms of the fluid flow and its gradients and from its definition is conserved:

$$V^\mu \partial_\mu \rho + \rho V^\mu{}_{;\mu} = 0 .$$

The interpretation though is that this density of "market stuff" per unit volume is the reciprocal of the volume element of the fluid per unit "market stuff". In physics theories, this "market stuff" is called mass. A second consequence is that if the internal energy is written as $\mu = \rho U$, then the internal energy U or energy density per unit "stuff" is to be considered as a general scalar function, not as a function of the density.

The modifications to the energy momentum tensor modify the field equations and the conservation laws for strategic-mass. To illustrate the additional behavior, I focus first on the change to the conservation law $V_\mu T^{\mu\nu}{}_{;\nu} = 0$ along the flow direction:

$$\mu_{;\nu} V^\nu + \mu V^\nu_{;\nu} + p V^\nu_{;\nu} = \eta \sigma_{\mu\nu} \sigma^{\mu\nu}$$

$$\Leftrightarrow V^\nu U_{;\nu} = -p V^\nu \left(\rho^{-1}\right)_{;\nu} + \rho^{-1} \eta \sigma_{\mu\nu} \sigma^{\mu\nu}.$$

The second form of the equation shows more explicitly the contributions to the rate of change to the internal energy per unit stuff $\mu\rho^{-1} = U$ defined as the energy per unit volume times the volume per unit stuff. The first term on the right represents the increase in internal energy due to the volume compression of the fluid and for a perfect fluid would be the only term; and the second term represents the increase in internal energy due to dissipation because of viscosity. I summarize the longitudinal conservation law:

$$\frac{D\mu}{\partial s} + (\mu + p) V^\mu{}_{;\mu} = \eta \sigma_{\mu\nu} \sigma^{\mu\nu}. \tag{A.9}$$

It is of course an open question as to whether there are viscous forces in a theory of games and if there are such forces whether they would obey the form as suggested here.

I consider next the transverse conservation laws $h_{\lambda\mu} T^{\mu\nu}{}_{;\nu} = 0$:

$$\begin{pmatrix} \left((\mu + p - \eta\theta) h^a_b - 2\eta\left(\sigma^a{}_b + \omega^a{}_b\right)\right) \dot{V}^b \\ + \left(-p_{;b} + \eta\left(1 - \tfrac{2}{n}\right)\theta_{;b}\right) h^{ab} + 2\eta_{;b}\sigma^{ab} + \eta h^{cd} h^{ab} V_{b;cd} \end{pmatrix} = 0. \tag{A.10}$$

The first bracket term is the acceleration with corrections due to viscosity. The second bracket term consists of the various transverse force terms: The pressure gradient contribution is the same as for a perfect fluid; next is a correction to the viscous force proportional to the shear and a standard viscous force equal to the divergence of the flow. Though these terms are complicated, they are present in real fluids and may play a role in realistic markets.

Appendix B

Symmetry in Differential Geometry

To make the symmetry idea more precise, I take machinery from differential geometry. I consider any two vector fields X^μ and Y^μ and the corresponding integral curves, each described by separate parameters:

$$\frac{dx^\mu}{ds} = X^\mu$$
$$\frac{dx^\mu}{dt} = Y^\mu$$

The change of the vector X^μ moving along the integral curve determined by vector Y^μ is given by the "Lie" derivative:

$$\mathbf{L}_\mathbf{Y}(\mathbf{X})^\mu = Y^\nu \partial_\nu X^\mu - X^\nu \partial_\nu Y^\mu = Y^\nu X^\mu{}_{;\nu} - X^\nu Y^\mu{}_{;\nu} . \tag{B.1}$$

There is always a frame in which the vector $\mathbf{Y}$ is constant. In this frame, only the first term in the first equation is non-zero; this is the derivative along $\mathbf{Y}$. Thus the Lie derivative provides the change of the vector X^μ along Y^μ. The result obtained is not covariant, since the first equation is not covariant. However, the Lie derivative can be expressed covariantly by the second equation. Because this expression is covariant, it holds in any frame. This is a proof of the assertion.

This derivative has a number of names in differential geometry: It is called the Lie derivative and the Lie product. The derivative is related to its function above in identifying variation along an integral curve. It is called a product in discussions of group theory.

It can be thought of as a product by considering the vector fields **X** and **Y** to be differential operators $\mathbf{X} = X^{\mu}\partial_{\mu}$ and $\mathbf{Y} = Y^{\mu}\partial_{\mu}$:

$$\mathbf{L_Y}(\mathbf{X}) = [\mathbf{X}, \mathbf{Y}]. \tag{B.2}$$

The Lie Derivative (Lie Product) is a new vector field, as evidenced by the second equality in Eq. (B.1). It is composed of tensor fields. The result is again a vector field and hence a new differential-operator. Thus it defines a product between two differential operators.

The Lie Derivative results from comparing vectors along two possible paths: Going along the curve s that defines the vector **X** and then along the curve t and going along the curve t that defines the vector **Y** and then along the curve s. The Lie Derivative provides the change of **X** along the integral curve defined by **Y**. The Lie Derivative can be extended to tensors. The expression for a second order tensor with lower indices is:

$$\mathbf{L}_Y(\mathbf{M})_{\mu\nu} = M_{\mu\nu;\lambda}Y^{\lambda} + M_{\lambda\nu}Y^{\lambda}{}_{;\mu} + M_{\mu\lambda}Y^{\lambda}{}_{;\nu}. \tag{B.3}$$

The Lie derivative determines a new tensor structure in terms of the old.

In particular it provides the change of a metric along a given integral curve. For a second order tensor, the Lie derivative can be written in terms of ordinary derivatives:

$$\mathbf{L}_Y(\mathbf{M})_{\mu\nu} = Y^{\lambda}\partial_{\lambda}M_{\mu\nu} + M_{\lambda\nu}\partial_{\mu}Y^{\lambda} + M_{\mu\lambda}\partial_{\nu}Y^{\lambda}.$$

The statement that the metric tensor does not change (has an ***isometry***) along a direction K^{μ} becomes the following statement which is valid in any frame: $\mathbf{L_K}(\mathbf{g}) = 0$. This can be evaluated using the above:

$$\mathbf{L_K}(\mathbf{g})_{\mu\nu} = K^{\lambda}g_{\mu\nu;\lambda} + g_{\lambda\nu}K^{\lambda}{}_{;\mu} + g_{\mu\lambda}K^{\lambda}{}_{;\nu} = g_{\lambda\nu}K^{\lambda}{}_{;\mu} + g_{\mu\lambda}K^{\lambda}{}_{;\nu} = 0.$$

I have used the fact that the covariant derivative of the metric vanishes. This says there is a frame in which the metric is independent of the coordinate; I say that the coordinate is ***inactive***.

The general result is that a coordinate is inactive if and only if there is a vector field satisfying the following condition:

$$K_{\mu;\nu} + K_{\nu;\mu} = 0. \tag{B.4}$$

Such vectors are called ***Killing vectors*** in the literature. I say such vectors form an ***inactive vector field***.

B.1 Metric Symmetry Group

The condition for an inactive choice provides the condition on the vector field given a metric. If the inactive vector field is given (such as positing a frame in which the vector field is a unit vector along the inactive choice), the condition can be turned around to yield a constraint on the metric:

$$g_{\mu\lambda}\partial_\nu K^\lambda + g_{\nu\lambda}\partial_\mu K^\lambda + K^\lambda \partial_\lambda g_{\mu\nu} = 0 . \tag{B.5}$$

For example, if the vector field is the unit vector $K^\mu = (0,\cdots,1,0,\cdots)$, the condition on the metric is that the derivative of the metric along that unit direction vanishes.

So far, I have considered the implications of the symmetry of the metric on the geodesic. If external sources are present, then the motion will depend on the energy momentum field. I now show that the symmetry of the metric extends to this field. Given an inactive vector field, there is a conserved ***current*** formed from the energy momentum field:

$$J^\mu = T^{\mu\nu} K_\nu . \tag{B.6}$$

I say a current is conserved if the covariant divergence of this current is zero:

$$J^\mu{}_{;\mu} = T^{\mu\nu}{}_{;\mu} K_\nu + T^{\mu\nu} K_{\nu;\mu} = 0 .$$

To each inactive vector field there will be a conserved current. The current is a generalization of the "charge" focusing on the sources.

For any two distinct inactive fields $\mathbf{K}_1$ and $\mathbf{K}_2$, the Lie product determines a new vector field $[\mathbf{K}_1,\mathbf{K}_2]$. It can be shown that this new vector field is itself an inactive field. This fact has wide application in physics [See for example Hawking and Ellis (1973)]. It means that the set of inactive fields along with the Lie product form a ***Lie Algebra***[43]:

$$[\mathbf{K}_\mathrm{i},\mathbf{K}_\mathrm{j}] = c_{ijk}\mathbf{K}_\mathrm{k} .$$

---

[43] I pass over the entire subject of local symmetry groups here. However, I believe that this subject will ultimately provide profound insight into the behavior of the theory. For future reference I provide two citations for the interested reader: For a physics-orientation see Hamermesh (1962); for a mathematical treatment see Chevalley (1946).

The coefficients c_{ijk} are structure constants that characterize the algebra. The elements of the algebra determine a Lie group and hence the (local) symmetry of the system. This symmetry is evidenced at each point as a local algebra of symmetries of the metric.

I illustrate the power of the symmetries for the example of the earth idealized as a sphere. Equation (3.1) is independent of longitude but I know there are additional symmetries. In fact there are three inactive vector fields:

$$\begin{aligned}\mathbf{K}_1 &= -\sin\phi\partial_\theta - \frac{\cos\phi}{\tan\theta}\partial_\phi \\ \mathbf{K}_2 &= -\cos\phi\partial_\theta + \frac{\sin\phi}{\tan\theta}\partial_\phi \,. \\ \mathbf{K}_3 &= \partial_\phi\end{aligned} \tag{B.7}$$

The choice of metric displays the invariance of rotations about the North South axis but does not display the other symmetries. The three inactive vector fields exhaust the symmetries. The Lie Product of each of any two of the inactive vectors is the third:

$$[\mathbf{K}_1, \mathbf{K}_2] = \mathbf{K}_3 \,. \tag{B.8}$$

This is the Lie Algebra that generates the rotation group $O(3)$. For a sphere, the group of rotations determines the metric. The fact that the Lie Product of any two inactive choices is not zero determines an important attribute of the line element. There is no frame in which the metric is independent of two coordinates. Thus Eq. (3.1) is independent of the longitude, ϕ but not the latitude θ.

B.2 Active and Inactive Geometries

I study the form the metric takes when there is a split of the coordinates between active and inactive. The split is similar to that for a sphere between the longitude and latitude components (inactive) and the radial components (active). For a sphere, the active components are absent since there is no variation along the radial direction.

But in higher dimensional spaces, there will in general be both types:

$$ds^2 = \gamma_{jk}\left(d\xi^j + A_a^j dx^a\right)\left(d\xi^k + A_b^k dx^b\right) + g_{ab}\, dx^a dx^b . \qquad \text{(B.9)}$$

The second term represents pure active components and has the general form Eq. (1.1).

In this appendix, I distinguish active choices with indices that are initial letters of the alphabet: $a,b,c,\cdots$. The coordinates are x^a. I refer to these coordinates collectively as forming the ***active geometry***. The first expression contains the inactive choices, denoted by indices $j,k,l,\cdots$. The coordinates are ξ^j. Collectively I denote these coordinates as the ***inactive geometry***.

The first term of Eq. (B.9) comprises both the inactive geometry and a mixture of active and inactive geometries. I recover the general form $\hat{\gamma}_{\mu\nu} dx^\mu dx^\nu$ by equating like-terms:

$$\begin{aligned} \hat{\gamma}_{jk} &= \gamma_{jk} \\ \hat{\gamma}_{ja} &= \gamma_{jk} A_a^k \\ \hat{\gamma}_{ab} &= g_{ab} + \gamma_{jk} A_a^j A_b^k \end{aligned} .$$

I call the vector field A_a^j a ***vector potential***. It represents the mixing between the active and inactive geometries. Each inactive coordinate has constraints given by Eq. (B.4):

$$\hat{\gamma}_{\mu\lambda}\partial_\nu K^\lambda + \hat{\gamma}_{\nu\lambda}\partial_\mu K^\lambda + K^\lambda \partial_\lambda \hat{\gamma}_{\mu\nu} = 0 .$$

I assert that I can go to a coordinate system in which the components of the Killing vector depend only on the inactive choices. Moreover, the only non-zero components of the Killing vector will be along the inactive choices:

$$K^a = 0, \quad \partial_a K^k = 0 .$$

This constrains the active and inactive components:

$$\begin{aligned} &\mathbf{K} \equiv K^k \partial_k \quad \mathbf{A_a} = A_a^j \partial_j \\ &\hat{\gamma}_{ak}\partial_b K^k + \hat{\gamma}_{bk}\partial_a K^k + \mathbf{K}\left(\hat{\gamma}_{ab}\right) = 0 \Rightarrow \mathbf{K}\left(\hat{\gamma}_{ab}\right) = 0 \\ &\hat{\gamma}_{ak}\partial_j K^k + \hat{\gamma}_{jk}\partial_a K^k + \mathbf{K}\left(\hat{\gamma}_{aj}\right) = 0 \Rightarrow \gamma_{kl} A_a^l \partial_j K^k + \mathbf{K}\left(\gamma_{jk} A_a^k\right) = 0 \\ &\gamma_{jl}\partial_k K^l + \gamma_{kl}\partial_j K^l + \mathbf{K}\left(\gamma_{jk}\right) = 0 \end{aligned}$$

The last equation is the condition on the metric in the inactive geometry. I rewrite these equations in a matrix form. I introduce a notation for the derivative of the Killing components: $U_j^{\ k} \equiv \partial_j K^k$. I start with the last equation, which allows determination of the inactive geometry from the components of the Killing vector:

$$\gamma U^T + U\gamma + \mathbf{K}\gamma = 0\,. \qquad \text{(B.10)}$$

I then use this equation to simplify the other two equations. For the active geometry, the result is:

$$\mathbf{K}\left(g_{ab}\right) = 0\,. \qquad \text{(B.11)}$$

Since this holds for the complete set of inactive vectors, it is a statement that the active geometry metric is independent of the inactive choices. The next equation is for the vector potential:

$$\left[\mathbf{K}, \mathbf{A_a}\right] = 0\,. \qquad \text{(B.12)}$$

In the inactive geometry, the Lie Product of the vector potential and the inactive vector vanishes.

In summary, I start with a set of inactive vectors along with their Lie Products $\left\{K_j^\mu\right\}$. I assert that I can find a special reference frame for the inactive vectors in this set such that the components of the inactive vectors depend only on a set of inactive choices $\left\{\xi^j\right\}$ and not on the active choices $\left\{x^a\right\}$. Moreover, I assert that I can always choose the coordinate system in such a way that the components of the Killing vectors are totally in the inactive space $X_j^a = 0$. By virtue of the symmetries, the metric will be greatly simplified.

Consider the properties of the metric $\hat{\gamma}_{\mu\nu}$ that result. There may in general be a set of inactive vectors that commute (*i.e.* whose Lie Product vanishes) with the full set of inactive vectors. Such vectors I call ***central***. Let us say one such vector is K_m^μ. It is then possible to choose the coordinate system in such a way that only a single component is non-zero and this component can be chosen to be unity: $X_m^m = 1$. This leads to the conclusion that both the active and inactive metric components will be independent of this variable. This is a significant simplification of the metric for central inactive vectors, as shown in Appendix C.

Appendix C

Central Strategies

If all inactive choices are central, the form of the line element will be given by Eq. (B.9) in a D dimensional space. Because all inactive vectors are central, all metric components are independent of the inactive choices. Since this specifies a specific frame, I call it the ***central frame***. It is then possible to evaluate the equations relating the metric to the sources, Eq. (1.2). I sketch the major steps but leave out some of the algebraic detail. I provide enough detail to get a sense of the arguments and allow the particular reader to be in a position to judge the merits of the argument.

I start by summarizing the result if all inactive vector fields are central. There will be three sets of equations: The curvature terms for the active geometry, the curvature terms for the inactive geometry and the curvature terms that are mixed.

The mixed terms are informative because they identify the conserved charged currents:

$$\begin{aligned} &\tfrac{1}{2}\frac{1}{\sqrt{|g\gamma|}}\partial_b\left(\sqrt{|g\gamma|}g^{ac}g^{bd}\gamma_{jk}F^k_{cd}\right)=\kappa\left(\mu+p\right)V_jV^a \\ &F^k_{cd}=\partial_c A^k_d-\partial_d A^k_c \\ &\gamma=\det\gamma_{jk} \\ &g=\det g_{ab} \end{aligned} \quad . \qquad (C.1)$$

The interpretation of this equation is interesting. Viewed in the active geometry, the vector potentials A^k_a determine an anti-symmetric tensor field F^j_{ab} in analogy with an electro–magnetic field. I call these ***market fields associated with the inactive component*** j. The source for the market field with the inactive component j is a current with a flow V^a, a "charge density" $2\kappa(\mu+p)V_j$ and a "charged" current $J_j^a = 2\kappa(\mu+p)V_j V^a$.

Source equations of this type can be viewed in many ways. For example, I may view it as applying to the case where the only inactive choices are the player value–choices. There is a vector potential associated with each inactive value–choice, ξ^j. It generates the player's "***market field***". In particular, in Appendix D I compute the player market field in terms of the Single Strategy connection [*Cf.* Eq. (D.3)]:

$$C_{\bar{j}m} = \frac{d\hat{\gamma}_{\bar{j}m}}{du} = \frac{d\left(\gamma_{\bar{j}\bar{k}} \bar{A}_m^{\bar{k}}\right)}{du} = \gamma_{\bar{j}\bar{k}} \frac{d\bar{A}_m^{\bar{k}}}{du} + C_{\bar{j}\bar{k}} \bar{A}_m^{\bar{k}} . \qquad \text{(C.2)}$$

$$C_{\bar{j}m} = \gamma_{\bar{j}\bar{k}} \bar{F}_{um}^{\bar{k}} + C_{\bar{j}\bar{k}} \bar{A}_m^{\bar{k}}$$

I use the "bar" notation to indicate that the implied summation goes over only the inactive value–choices $\bar{k}$ and not over the inactive time and strategy choices m. For each player I work in the specific gauge $\bar{A}_u^{\bar{k}} = 0$. One possible boundary condition for the Single Strategy Game is a common defensive position for each player: $\bar{F}_{um}^{\bar{k}} V^m = 0$. In the co-moving frame, this requires the time component of the market field $\bar{F}_{u0}^{\bar{k}}$ to vanish.

Returning now to the general exposition, in the active geometry, the inactive geometry components γ_{jk} appear as scalar fields:

$$\frac{1}{\sqrt{|g\gamma|}} \gamma_{ik} \partial_a \left(\sqrt{|g\gamma|} g^{ab} \gamma^{kl} \partial_b \gamma_{lj} \right) = \begin{pmatrix} \frac{1}{2} \gamma_{ik} \gamma_{jl} F_{ac}^k F_{bd}^l g^{ab} g^{cd} \\ -2\kappa \begin{pmatrix} (\mu+p) V_i V_j \\ -\dfrac{\mu-p}{D-2} \gamma_{ij} \end{pmatrix} \end{pmatrix} . \qquad \text{(C.3)}$$

These are second order equations for the scalar fields where the sources are the scalar fields themselves, the market fields and the charges.

In the active geometry, the active components appear similar to the original Eqs. (1.2) with additional sources from the scalar and market fields:

$$\begin{pmatrix} R_{ab} + \frac{1}{2} \gamma_{jk} F_{ac}^j F_{bd}^k g^{cd} \\ + \frac{1}{2} \gamma^{jk} \gamma_{jk;ab} + \frac{1}{4} \partial_a \gamma^{jk} \partial_b \gamma_{jk} \end{pmatrix} = -\kappa \left((\mu+p) g_{ac} g_{bd} V^c V^d - \frac{\mu-p}{D-2} g_{ab} \right) .$$

The conserved tensor is the active geometry "energy–momentum" tensor that is computed from the curvature:

$$R_{ab} - \tfrac{1}{2} g_{ab} R = \begin{pmatrix} -\kappa\left(\left(\mu+p\right) g_{ac} g_{bd} V^c V^d - p g_{ab}\right) \\ -\tfrac{1}{2}\gamma_{jk} F^j_{ac} F^k_{bd} g^{cd} + \tfrac{1}{8} g_{ab} g^{qc} g^{pd} \gamma_{mn} F^m_{pc} F^n_{dq} \\ -\tfrac{1}{2}\gamma^{jk}\gamma_{jk;ab} - \tfrac{1}{4}\partial_a \gamma^{jk}\partial_b \gamma_{jk} \\ +\tfrac{1}{2} g_{ab}\left(\gamma^{jk} g^{cd}\gamma_{jk;cd} + \tfrac{1}{4} g^{cd}\gamma^{mn}\gamma^{jk}\gamma_{mn;c}\gamma_{jk;d} + \tfrac{3}{4} g^{cd}\gamma^{jk}{}_{;c}\gamma_{jk;d}\right) \end{pmatrix} . \quad \text{(C.4)}$$

As before, these three equations obey Eq. (3.8). The source behavior can be articulated in terms of the active and inactive geometries. First, the conservation of strategic-mass takes the form:

$$\partial_a\left(\rho\sqrt{\gamma g} V^a\right) = 0 . \quad \text{(C.5)}$$

The inactive flow vectors define a set of conserved "charges" that follow from Eq. (C.1):

$$\frac{dV_j}{ds} + \frac{1}{\mu+p}\frac{dp}{ds} V_j = 0 \Rightarrow \frac{d}{ds}\left(\frac{\mu+p}{\rho} V_j\right) = 0$$

$$c_j \equiv \frac{\mu+p}{\rho} V_j \Rightarrow \frac{dc_j}{ds} = 0$$

Along a streamline, the ***charge*** c_j is constant. This shows that the current $J^a_j = 2\kappa\left(\mu+p\right) V_j V^a = 2\kappa\rho c_j V^a$ is proportional to the market density current and hence conserved.

I convert the expression for the charge into an equation for the inactive choice:

$$V_j = \hat{\gamma}_{j\mu} V^\mu = \hat{\gamma}_{jk} V^k + \hat{\gamma}_{ja} V^a = \gamma_{jk} V^k + \gamma_{jk} A^k_a V^a = \frac{\rho}{\mu+p} c_j$$

$$\gamma_{jk}\frac{d\xi^k}{ds} + \gamma_{jk} A^k_a \frac{dx^a}{ds} = \frac{\rho}{\mu+p} c_j$$

$$\frac{d\xi^j}{ds} + A^j_a \frac{dx^a}{ds} = \frac{\rho}{\mu+p}\gamma^{jk} c_k$$

Because the charge is conserved I obtain:

$$\frac{dV_j}{ds} + \frac{1}{\mu + p}\frac{dp}{ds}V_j = 0. \tag{C.6}$$

Therefore I get the significant result that the inactive charge flow is determined by the pressure and energy density.

Given a solution to the above set of equations, a class of additional solutions can be generated making the "gauge substitutions":

$$\begin{aligned} &\xi^j \to \xi^j + \Lambda^j \\ &V^j \to V^j + V^a \partial_a \Lambda^j . \\ &A_a^j \to A_a^j - \partial_a \Lambda^j \end{aligned} \tag{C.7}$$

The full line element is invariant under these transformations where the active metric g_{ab} and the inactive metric γ_{jk} are held constant under the above gauge transformation. The solution for the inactive choice flow is exactly what one would expect for a conserved momentum along the flow line.

The active component can be expressed in a gauge invariant fashion, again starting from Eq. (3.8):

$$V^a{}_{;b} V^b + \tfrac{1}{2} g^{ab} V_j V_k \partial_b \gamma^{jk} - V_k g^{ab} F^k_{bc} V^c - \left(g^{ab} - V^a V^b\right)\frac{\partial_b p}{\mu + p} = 0. \tag{C.8}$$

These flow equations are direct consequences of the field equations. They result from identities that hold for the curvature.

Because of the original source equation, the normalization of the flow vector is constant along the streamlines $d\left(\hat{\gamma}_{\mu\nu}V^\mu V^\nu\right)/ds = 0$:

$$\frac{d}{ds}\left(\hat{\gamma}_{\mu\nu}V^\mu V^\nu\right) = 2\hat{\gamma}_{\mu\nu}V^\mu \frac{DV^\nu}{\partial s} = 2\hat{\gamma}_{\mu\nu}V^\mu\left(\hat{\gamma}^{\nu\lambda} - V^\nu V^\lambda\right)\frac{\partial_\lambda p}{\mu + p}.$$

$$\frac{d}{ds}\left(\hat{\gamma}_{\mu\nu}V^\mu V^\nu\right) = 2\left(1 - \hat{\gamma}_{\mu\nu}V^\mu V^\nu\right)\frac{1}{\mu + p}\frac{dp}{ds}$$

I see that if the initial condition is made that the flow is a unit vector, then the flow remains a unit vector:

$$\gamma^{jk}V_j V_k + g_{ab}V^a V^b = 1. \tag{C.9}$$

I use the fact that the normalization can be expressed in a gauge invariant fashion showing the contribution to the normalization from the charges.

To solve these equations, attention must be given to the fact that the number of equations exceeds the number of variables: Given any solution, another solution is obtained by an arbitrary diffeomorphism. I have used the concept of a gauge transformation above; I now define the concept in more detail.

It is standard to choose a "gauge condition" to eliminate the additional degrees of freedom [*Cf.* Hawking and Ellis (1973)]. In this way, it can be shown that there are solutions to the equations; in particular there are weak field approximations to the equations. The standard or ***harmonic gauge*** conditions are defined to be:

$$\psi^{\mu} \equiv \hat{\gamma}^{\mu\nu}_{|\nu} - \tfrac{1}{2}\hat{\gamma}^{\mu\nu}\hat{\gamma}_{\lambda\rho}\delta g^{\lambda\rho}{}_{|\nu} = 0 . \qquad \text{(C.10)}$$

Hawking and Ellis give a good discussion, including defining the above gauge condition according to a "background" metric. In such an instance, the covariant connection that appears above is relative to the "background" metric.

When all inactive choices are central, there will be gauge conditions for both the active and inactive components. The inactive components satisfy:

$$\left\{ \begin{aligned} &\frac{1}{\sqrt{g\gamma}}\left(\sqrt{g\gamma}\, g^{ab}\right)_{||b} - \tilde{g}^{ac} g^{bd}\,\delta A^{k}_{d}\,\tilde{\gamma}_{jk}\,\tilde{F}^{j}_{bc} \\ &-\tfrac{1}{2}\left(\gamma^{jk} + g^{cd}\,\delta A^{j}_{c}\,\delta A^{k}_{d}\right)\tilde{g}^{ab}\,\partial_{b}\tilde{\gamma}_{jk} \end{aligned} \right\} = 0 . \qquad \text{(C.11)}$$

The gauge conditions depend on the background market fields $\tilde{F}^{j}_{bc}$ and the background scalar fields $\tilde{\gamma}_{jk}$. The difference between the solution to the field equations and the background field is indicated by the δ: For example the difference between the market potential and the background market potential is $\delta A^{j}_{a} = A^{j}_{a} - \tilde{A}^{j}_{a}$. The corresponding gauge conditions for the inactive components are:

$$\frac{1}{\sqrt{g\gamma}}\left(\sqrt{g\gamma}\, g^{ab}\,\tilde{\gamma}_{kj}\,\delta A^{j}_{a}\right)_{||b} = 0 . \qquad \text{(C.12)}$$

In the simplest case where the background metric components are constant, the gauge conditions are:

$$\frac{1}{\sqrt{g\gamma}}\partial_b\left(\sqrt{g\gamma}g^{ab}\right)=0$$
$$\frac{1}{\sqrt{g\gamma}}\partial_b\left(\sqrt{g\gamma}g^{ab}A_a^j\right)=0$$

The weak field approximation assumes that the source occurs with a weak coupling κ and only the lowest order terms need to be kept:

$$\begin{aligned}
&\tfrac{1}{2}\tilde{g}^{ab}\partial_a\partial_b\gamma^{jk}\cong\kappa\left((\mu+p)\tilde{\gamma}^{ij}\tilde{\gamma}^{kl}V_iV_l-\frac{\mu-p}{D-2}\tilde{\gamma}^{jk}\right)\\
&-\tfrac{1}{2}\tilde{g}^{cd}\partial_c\partial_dA_a^j\cong\kappa(\mu+p)\tilde{\gamma}^{jk}\tilde{g}_{ab}V_kV^b\\
&\left\{\begin{aligned}&\tfrac{1}{2}\tilde{g}^{cd}\partial_c\partial_d\left(\delta g^{ab}-\tfrac{1}{2}\tilde{g}^{ab}\tilde{g}_{cd}\delta g^{cd}-\tfrac{1}{2}\tilde{g}^{ab}\tilde{\gamma}_{jk}\delta\gamma^{jk}\right)\\&=\kappa\left((\mu+p)V^aV^b-p\tilde{g}^{ab}\right)\end{aligned}\right\}
\end{aligned}\quad. \tag{C.13}$$

To the same approximation, the conservation laws are:

$$\begin{aligned}
&\partial_a\left(\rho V^a\right)-\tfrac{1}{2}\rho V^a\tilde{\gamma}_{jk}\partial_a\delta\gamma^{jk}-\tfrac{1}{2}\rho V^a\tilde{g}_{cd}\partial_a\delta g^{cd}=0\\
&V^a\partial_aV_j+\frac{V^a\partial_ap}{\mu+p}V_j=0\\
&\left\{\begin{aligned}&\tilde{g}_{ac}V^c{}_{;e}V^e-V_j\delta F^j_{ac}V^c+\tfrac{1}{2}V_jV_k\delta\gamma^{jk}{}_{;a}\\&-\left(\delta_a^e-\tilde{g}_{ac}V^cV^e\right)\frac{p_{;e}}{\mu+p}\end{aligned}\right\}=0
\end{aligned}\quad. \tag{C.14}$$

The gauge conditions for the active components are:

$$\partial_a\left(\delta g^{ab}-\tfrac{1}{2}\tilde{g}^{ab}\tilde{g}_{cd}\delta g^{cd}-\tfrac{1}{2}\tilde{g}^{ab}\tilde{\gamma}_{jk}\delta\gamma^{jk}\right)=0. \tag{C.15}$$

The power of the weak field approximation is that the solutions are standard: $\tilde{g}^{ab}\partial_a\partial_b\phi=source$. This might be important when looking for numerical solutions.

Appendix D

Single Strategy Model

I specialize the discussion "all inactive strategies are central with some number of active strategies" to the special case of a single active strategy. I take time–choice t to be inactive, the value–choices to be inactive and all strategy–choices to be inactive but one, the active strategy R. The full set of equations simplifies:

$$\begin{aligned}
&\tfrac{1}{2}\frac{1}{\sqrt{|g\gamma|}}\partial_R\left(\sqrt{|g\gamma|}g^{RR}g^{RR}\gamma_{jk}F^k_{RR}\right)=\kappa(\mu+p)V_jV^R\\
&g_{RR}V^RV^R+\gamma^{jk}V_jV_k=1\\
&g=g_{RR} \qquad\qquad\qquad\qquad . \qquad (D.1)\\
&\tfrac{1}{2}\frac{1}{\sqrt{|g\gamma|}}\partial_R\left(\sqrt{|g\gamma|}g^{RR}\gamma^{kl}\partial_R\gamma_{lj}\right)=-\kappa\begin{pmatrix}(\mu+p)\gamma^{kl}V_lV_j\\ -\frac{\mu-p}{D-2}\delta^k_j\end{pmatrix}\\
&R_{RR}+\tfrac{1}{2}\gamma^{jk}\gamma_{jk;RR}+\tfrac{1}{4}\partial_R\gamma^{jk}\partial_R\gamma_{jk}=\kappa\frac{\mu-p}{D-2}g_{RR}
\end{aligned}$$

From the first equation it follows that the flow along the active direction is zero: $V^R=0$, showing that the direction R is transverse to the flow. The flow is zero because the antisymmetric matrix F^k_{RR} necessarily vanishes independent of the market potentials A^k_R for each component k. The remaining components of flow determine a time-like vector because of the normalization: $\gamma_{jk}V^jV^k=1$.

D.1 Solution

The equations can be written in a simpler form using the following definitions:

$$\begin{aligned}
\mu &\equiv \mu_0 \left|\frac{\gamma_0}{\gamma}\right| \\
p &\equiv p_0 \left|\frac{\gamma_0}{\gamma}\right| \\
C_{jk} &\equiv \sqrt{\left|\frac{\gamma}{g}\right|} \frac{1}{\sqrt{|\gamma_0|}} \frac{d\gamma_{jk}}{dR} \\
\frac{du}{dR} &\equiv \sqrt{|\gamma_0|} \sqrt{\left|\frac{g}{\gamma}\right|}
\end{aligned} \tag{D.2}$$

The gauge condition Eq. (C.11) determines the active metric g_{RR}:

$$\begin{aligned}
&\frac{d}{dR}\left(\sqrt{g\gamma g^{-1}}\right) = 0 \\
&\frac{d}{dR}\left(\sqrt{\frac{\gamma}{g}}\right) = 0 \Rightarrow \frac{du}{dR} = \text{constant}
\end{aligned}$$

The "standard" gauge condition is that the active strategy distance is given by the above reduced variable.

To recover the property that strategy–choices are non-negative, I may need to make a change of variables after the solution is obtained. An example might be:

$$y = e^u .$$

This new variable y is non-negative and is determined by the natural variable u.

I introduce a reduced pressure and energy density, a connection C_{jk} and a ***natural distance*** for the active strategy so that the active components contribute $-\gamma du^2$ to the line element.

The resultant equations are coupled linear first order differential equations in the metric and connection:

$$\frac{d\gamma_{jk}}{du} = C_{jk} = C^l_j \gamma_{lk}$$
$$\frac{dC^k_j}{du} = 2\kappa\left(\left(\mu_0 + p_0\right)V^k V_j - \frac{\mu_0 - p_0}{D-2}\delta^k_j\right). \qquad \text{(D.3)}$$
$$C^k_l C^l_k - C^j_j C^l_l = -8\kappa p_0$$

The last equation is an algebraic (as opposed to a differential) equation and results from the active geometry equation. It shows that the reduced pressure is determined. I assume that the reduced energy density is a function of the reduced pressure.

The time-like flow vector will be a function of the active choice only. The general form of the charge gradients is [*Cf.* Appendix F]:

$$V_{\mu;\nu} = \dot{V}_\mu V_\nu + \theta_{\mu\nu} + \omega_{\mu\nu}.$$

The vorticity $\omega_{\mu\nu}$ and expansion $\theta_{\mu\nu}$ parameters are orthogonal to the flow. The field equations determine the acceleration in terms of the transverse gradient of the pressure:

$$\dot{V}^u \equiv \Gamma^u_{jk} V^j V^k = \tfrac{1}{2}\gamma^{-1}\frac{d\gamma_{jk}}{du}V^j V^k = -\gamma^{-1}\frac{1}{\mu+p}\frac{dp}{du}.$$

To better articulate the possibilities, I introduce a full set of orthonormal vectors U^j_p that include the flow vector $V^j \equiv U^j_0$. For a path transverse to the streamline along the active strategy direction starting from some fixed point, the orthonormal set of vectors $\mathbf{U}_p = U^j_p$ defines a ***transverse evolution frame*** that changes continuously with the active strategy value:

$$\mathbf{U}_p = \mathbf{B}_{pq}\mathbf{U}_q(0).$$

I can specify the possibilities by specifying the conditions on the ***transverse evolution matrix*** $\mathbf{B}$. The condition that the vectors are orthonormal is that:

$$\mathbf{B}\overline{\boldsymbol{\gamma}}\mathbf{B}^{\mathrm{T}} = \overline{\mathbf{I}} \equiv \begin{pmatrix} 1 & 0 \\ 0 & -\delta_{\alpha\beta} \end{pmatrix}.$$

The transverse directions are normalized with a minus sign compared to the time like direction along the flow. The Greek indices are used to indicate the transverse directions when a distinction is needed. The orthogonality condition is specified in terms of the metric $\bar{\boldsymbol{\gamma}}$ in the new basis. The completeness relation follows from the orthogonality:

$$\mathbf{B}^{\mathrm{T}}\bar{\mathbf{I}}\mathbf{B} = \bar{\boldsymbol{\gamma}}^{-1}.$$

As one moves along the active strategy direction from some fixed point, the gradient of the orthonormal vectors change and can be expressed in terms of the orthonormal basis:

$$\frac{d\mathbf{B}}{du}\bar{\boldsymbol{\gamma}}\mathbf{B}^{T} = \mathbf{f} = \begin{pmatrix} -\tfrac{1}{2}f & 2\theta_{\alpha} \\ -2\omega_{\alpha} & \left(-\tfrac{1}{2}c_{\alpha\beta} + b_{\alpha\beta}\right) \end{pmatrix}.$$

Using the completeness relationship, this can be written as a linear differential equation for the transverse evolution matrix:

$$\frac{d\mathbf{B}}{du} = \mathbf{f}\bar{\mathbf{I}}\mathbf{B}$$
$$\frac{d\bar{\boldsymbol{\gamma}}}{du} = \bar{\mathbf{C}}$$

I have included the related evolution equation for the metric which determines the connection in the transverse evolution basis.

There is a relationship between the transverse evolution connection and the transverse evolution matrix based on the fact that the normalization condition must be satisfied along the path:

$$\frac{d}{du}\mathbf{B}\bar{\boldsymbol{\gamma}}\mathbf{B}^{\mathrm{T}} = \frac{d\bar{\mathbf{I}}}{du} = 0$$
$$\frac{d\mathbf{B}}{du}\bar{\boldsymbol{\gamma}}\mathbf{B}^{\mathrm{T}} + \mathbf{B}\bar{\boldsymbol{\gamma}}\frac{d\mathbf{B}^{\mathrm{T}}}{du} + \mathbf{B}\frac{d\bar{\boldsymbol{\gamma}}}{du}\mathbf{B}^{\mathrm{T}} = 0.$$
$$\mathbf{f} + \mathbf{f}^{T} + \mathbf{B}\bar{\mathbf{C}}\mathbf{B}^{\mathrm{T}} = 0$$

This demonstrates that the transverse evolution connection as expressed in terms of the matrix elements of the orthonormal set is determined by the generators $\mathbf{f}$.

The evolution is determined by the connection in a transformed basis and is determined by the vorticity, compression and gravitational force:

$$\mathbf{c} \equiv \mathbf{B}\bar{\mathbf{C}}\mathbf{B}^{\mathrm{T}} = \begin{pmatrix} f & 2(\omega_\alpha - \theta_\alpha) \\ 2(\omega_\alpha - \theta_\alpha) & c_{\alpha\beta} \end{pmatrix}.$$

I can now relate the elements that appear in the transverse evolution connection $\mathbf{c}$ to attributes of the flow and connection. The time-time component of the transverse evolution connection is the invariant gravitational force $f = C_{jk} V^j V^k$. This item determines the length of the flow e^{-v} when viewed in the transverse evolution frame: $f = 2dv/du$.

The decomposition of the flow vector into the vorticity and expansion parameters leads to the following determination:

$$\begin{aligned}
&V_{u;u} = -\Gamma^l_{uu} V_l = 0 = \theta_{uu} \\
&V_{j;k} = -\Gamma^l_{jk} V_l = \omega_{jk} + \theta_{jk} + \dot{V}_j V_k = \omega_{jk} + \theta_{jk} \Rightarrow \omega_{jk} = \theta_{jk} = 0 \\
&V^j{}_{;u} = \frac{dV^j}{du} + \tfrac{1}{2} C^j{}_k V^k = \omega^j{}_u + \theta^j{}_u \\
&V_{u;k} = -\Gamma^l_{uk} V_l = -\tfrac{1}{2} C_{km} V^m = \omega_{uk} + \theta_{uk} + \dot{V}_u V_k \\
&V_{u;k} = -\omega_{ku} + \theta_{ku} - \tfrac{1}{2} C_{mn} V^m V^n V_k \\
&\therefore \omega^j{}_u \equiv -\omega_\alpha U^j_\alpha \quad \theta^j{}_u \equiv -\theta_\alpha U^j_\alpha
\end{aligned}$$

The parameters that appear in the transverse evolution equation are in fact the projections of the vorticity and expansion along the transverse evolution basis vectors.

As a consequence of the decomposition into the transverse evolution basis, the connection and metric in the original basis are determined:

$$\begin{aligned}
&\gamma^{jk} = V^j V^k - U^j_\alpha U^k_\beta \\
&C^{jk} = f V^j V^k - 2(\omega_\alpha - \theta_\alpha)(U^j_\alpha V^k + V^j U^k_\alpha) + c_{\alpha\beta} U^j_\alpha U^k_\beta
\end{aligned}$$

It remains of course to determine the transverse evolution connection components that appear.

The parameters that appear in the decomposition of the transverse evolution connection are related to:

- A ***transverse torsion*** $b_{\alpha\beta}$;
- A ***transverse expansion*** $\Theta = -c_{\alpha\alpha}$ and
- A ***transverse shear*** $\Sigma_{\alpha\beta} = -c_{\alpha\beta} + c_{\gamma\gamma}\delta_{\alpha\beta}/(D-2)$.

The choice of basis in which the vector fields are eigenvectors determines a unit cell with orthogonal axes. This cell changes shape for a differential change along the active strategy according to the transverse expansion, shear and torsion.

I further illuminate the concept of transverse expansion by looking at the trace of the connection, which determines the overall expansion along the transverse active strategy:

$$\gamma^{jk}C_{jk} = \frac{d\ln\gamma}{du} = f - c_{\gamma\gamma} = f + \Theta .$$

The equation describes the change in shape of the full geometry as a sum of the change in length along the flow direction and the transverse change in geometry Θ.

I solve the equations of motion Eq. (D.3) by expressing them in terms of the above parameterization:

$$\mathbf{s} = 2\kappa \begin{pmatrix} \mu_0 + p_0 - \dfrac{\mu_0 - p_0}{D-2} & 0 \\ 0 & \dfrac{\mu_0 - p_0}{D-2}\delta_{\alpha\beta} \end{pmatrix} = \mathbf{B}\frac{d\bar{\mathbf{C}}}{du}\mathbf{B}^{\mathrm{T}} - \mathbf{B}\bar{\mathbf{C}}\bar{\boldsymbol{\gamma}}^{-1}\bar{\mathbf{C}}\mathbf{B}^{\mathrm{T}} .$$

The left hand side is determined by the sources; the right hand side is determined by the parameters of the model that include the metric and flow parameters. The right hand side can be simplified using the definitions, which results in an equation for the scalar parameters:

$$\begin{aligned} \frac{d\mathbf{c}}{du} &= \frac{d\mathbf{B}}{du}\bar{\mathbf{C}}\mathbf{B}^T + \mathbf{B}\frac{d\bar{\mathbf{C}}}{du}\mathbf{B}^T + \mathbf{B}\bar{\mathbf{C}}\frac{d\mathbf{B}^T}{du} \\ \frac{d\mathbf{c}}{du} &= \mathbf{s} + \mathbf{f}^T\bar{\mathbf{I}}\mathbf{f} - \bar{\mathbf{f}}\bar{\mathbf{I}}\mathbf{f}^T \end{aligned} .$$

These equations can be converted back to the scalar notation that allows the various contributions to be seen:

$$\frac{df}{du} = 2\kappa\left(\mu_0 + p_0 - \frac{\mu_0 - p_0}{D-2}\right) - 4\omega_\alpha\omega_\alpha + 4\theta_\alpha\theta_\alpha$$

$$\frac{d(\omega_\alpha - \theta_\alpha)}{du} = \left\{\begin{array}{l} \tfrac{1}{2}\left(\frac{\Theta}{D-2} - f\right)(\theta_\alpha + \omega_\alpha) \\ +\tfrac{1}{2}\Sigma_{\alpha\beta}(\omega_\beta + \theta_\beta) - b_{\alpha\beta}(\omega_\beta - \theta_\beta) \end{array}\right\}$$

$$\frac{d\Sigma_{\alpha\beta}}{du} = \left\{\begin{array}{l} 4\omega_\alpha\omega_\beta - 4\theta_\alpha\theta_\beta - \frac{4\omega_\gamma\omega_\gamma - 4\theta_\gamma\theta_\gamma + 2\Sigma_{\delta\gamma}b_{\gamma\delta}}{D-2}\delta_{\alpha\beta} \\ +\Sigma_{\alpha\gamma}b_{\gamma\beta} - b_{\alpha\gamma}\Sigma_{\gamma\beta} \end{array}\right\}. \quad \text{(D.4)}$$

$$\frac{d\Theta}{du} = -2\kappa(\mu_0 - p_0) + 4\omega_\alpha\omega_\alpha - 4\theta_\alpha\theta_\alpha + 2\Sigma_{\alpha\beta}b_{\beta\alpha}$$

$$2\kappa p_0 = \left\{\begin{array}{l} \tfrac{1}{2}f\Theta + \tfrac{1}{4}\frac{D-3}{D-2}\Theta^2 - \tfrac{1}{4}\Sigma_{\alpha\beta}\Sigma_{\alpha\beta} \\ +(\omega_\alpha - \theta_\alpha)(\omega_\alpha - \theta_\alpha) \end{array}\right\}$$

I have more parameters here than equations and so additional conditions can be added to make the solutions well determined. I make some tentative choices in what follows (primarily for numerical models) though for certain arguments I will return to the general solution.

One simplifying assumption is to set the transverse torsion to zero. The antisymmetric matrix $b_{\alpha\beta}$ describes the transverse rotations of the cell that occur for a motion along the active strategy and that lie entirely in the inactive space. The antisymmetric matrix $b_{\alpha\beta}$ is arbitrary and there are no differential equations that constrain it. At every point I can choose an arbitrary set of vectors as an orthonormal set. I take the flow vector as one such unit time-like vector and pick any set of orthonormal vectors U_α^j that are orthogonal to the flow. The number of such transverse orthonormal vectors is $\tfrac{1}{2}N(N-1)$ where the number of transverse dimensions is $N = D-2$. This is the same number of unknown components of the antisymmetric matrix $b_{\alpha\beta}$. Thus these components are manifestations of the fact that the orthonormal set of transverse vectors can be oriented arbitrarily. I choose the antisymmetric matrix to vanish, which is equivalent to choosing a basis in which there is no transverse torsion. The transverse evolution frame is fixed.

I still have more unknowns than equations. I introduce the ***centrally co-moving hypothesis*** as the statement that the transverse evolution frame is also central. The fundamental assumption is that in the central frame, the metric and flow components are independent of time. I believe that this hypothesis is reasonable given that the total flow is orthogonal to the active direction along which the flow is always zero. The consequence of this is that the expansion parameters θ_α are zero. I now prove this result.

In the central frame, the metric components are independent of the inactive choices in general and the time–choice in particular. This implies the existence of a time-like vector K^j whose only component is along the time direction. It follows that this vector satisfies the relation:

$$K_{\mu;\nu} + K_{\nu;\mu} = 0 .$$

If the transverse evolution frame is also a central frame, then since it has a unit flow defined along the time direction, the two vectors are proportional in all frames:

$$K^\mu = \phi V^\mu .$$

This puts constraints on both the proportionality scalar ϕ and the flow gradient $V_{\mu;\nu} = \dot{V}_\mu V_\nu + \theta_{\mu\nu} + \omega_{\mu\nu}$:

$$\phi_{;\mu} = -\phi \dot{V}_\mu$$
$$K_{\mu;\nu} = \phi_{;\nu} V_\mu - \phi_{;\mu} V_\nu + \phi\theta_{\mu\nu} + \phi\omega_{\mu\nu} = -K_{\nu;\mu} .$$
$$\therefore \theta_{\mu\nu} = 0$$

The centrally co-moving hypothesis is equivalent to requiring the expansion coefficients to be zero. One consequence of the centrally co-moving hypothesis is that the variation of the flow with distance is particularly simple:

$$\frac{dV^j}{du} = -\tfrac{1}{2} f V^j .$$

In particular, if the initial conditions start with a space component of the flow to be zero, it remains zero. I exclude consideration of the time-component as zero since the flow is assumed to be a time-like vector. Note that the vorticity coefficients are still free to be non-zero. The centrally co-moving hypothesis does not constrain the size of the vorticity, though the field equations do.

The resultant set of field equations can be recast in terms of these transverse evolution connection components, using the geometric notation. The metric and connection are then obtained from them:

$$
\begin{aligned}
&\frac{d\ln\gamma}{du} = \Theta + f \\
&\frac{d\omega_\alpha}{du} = \left(\tfrac{1}{2}\left(-f + \frac{\Theta}{D-2} \right)\delta_{\alpha\beta} + \tfrac{1}{2}\Sigma_{\alpha\beta} \right)\omega_\beta \\
&\frac{d\Theta}{du} = -2\kappa\left(\mu_0 - p_0\right) + 4\omega_\alpha\omega_\alpha \\
&\frac{d\Sigma_{\alpha\beta}}{du} = 4\omega_\alpha\omega_\beta - \frac{4\omega_\gamma\omega_\gamma}{D-2}\delta_{\alpha\beta} \\
&\frac{df}{du} = 2\kappa\left(\mu_0 + p_0 - \frac{\mu_0 - p_0}{D-2} \right) - 4\omega_\alpha\omega_\alpha \\
&2\kappa p_0 = \tfrac{1}{2} f\Theta + \tfrac{1}{4}\frac{D-3}{D-2}\Theta^2 - \tfrac{1}{4}Tr\boldsymbol{\Sigma}^2 + 2\omega_\alpha\omega_\alpha \\
&\frac{dV^k}{du} = -\tfrac{1}{2} fV^k \\
&\frac{dU_\alpha^j}{du} = -2\omega_\alpha V^j - \tfrac{1}{2}\Sigma_{\alpha\beta}U_\beta^j - \tfrac{1}{2}\frac{\Theta}{D-2}U_\alpha^j \\
&\gamma^{jk} = V^j V^k - U_\alpha^j U_\alpha^k \\
&C^{jk} = \begin{Bmatrix} -U_\alpha^j \Sigma_{\alpha\beta} U_\beta^k + \dfrac{\Theta}{D-2}\left(\gamma^{jk} - V^j V^k\right) + fV^j V^k \\ -2\omega_\alpha U_\alpha^j V^k - 2\omega_\alpha U_\alpha^k V^j \end{Bmatrix}.
\end{aligned}
\qquad \text{(D.5)}
$$

There is now one more scalar unknown than there are equations. I can reduce the number of unknown scalars in a variety of ways by relating the reduced energy to the reduced pressure. For example I can choose the reduced energy to be a constant in analogy to real fluids that are ***incompressible*** or I can take the reduced energy to be proportional to the reduced pressure in analogy to real fluids that are ***homogeneous***. For now I assume only that some such choice will be made and that the equations have a unique solution given a set of initial conditions to specify each scalar function.

The last four equations in Eq. (D.5) can be used to evaluate the flow and transverse flow vectors, the metric and the connection. The listed algebraic pressure equation is to be used in the differential scalar equations for the remaining variables. The scalar equations provide unique solutions from any given set of initial conditions.

D.2 Scalar Properties

I make the following observations about the meaning of the scalars. Because there is no flow along an active choice, the conservation of strategic-mass, Eq. (C.5), is satisfied trivially. More generally, the scalars are constant along the streamline; the distance u is a fixed constant on the streamline. The differential field equations therefore define the behavior transverse to the streamline that lies along the active strategy.

One consequence is that the acceleration of the inactive strategy is determined by the market gauge and will normally be set to zero:

$$\dot{V}^j \equiv V^j{}_{;k} V^k = \hat{\gamma}^{ju} \frac{1}{\mu + p} \frac{dp}{du} = \gamma^{-1} A^j_u \frac{1}{\mu + p} \frac{dp}{du}.$$

For the active component, there is acceleration and I recover Eq. (C.8):

$$\dot{V}^u \equiv \Gamma^u_{jk} V^j V^k = -\gamma^{-1} \frac{1}{\mu + p} \frac{dp}{du} \Rightarrow$$

$$\frac{1}{\mu + p} \frac{dp}{du} = -\tfrac{1}{2} \frac{d\gamma_{jk}}{du} V^j V^k = -\tfrac{1}{2} f$$

There is acceleration along the transverse direction, which is matched by the gradient of the pressure. I see that f is a measure of that acceleration, which is the analog of the force concept in physics.

The centrally co-moving hypothesis leads to the conclusion that the inactive vector associated with time is equal to the flow vector times a scalar function: $K^\mu = \phi V^\mu$. There is a general argument [for example see

Hawking and Ellis (1973)] that shows that this scalar function determines a ***gravitational field***. The argument goes as follows. I start with the form of the gradient $K_{\mu;\nu}$ including vorticity as a contribution and obtain the expression for the gradient of the scalar in terms of the acceleration:

$$K_{\nu;\mu} = \phi_{;\mu} V_\nu + \phi V_{\nu;\mu}$$

$$0 = V^\mu V^\nu K_{\mu;\nu} = \phi_{;\mu} V^\mu = \frac{d\phi}{ds}$$

$$V^\nu K_{\nu;\mu} = \phi_{;\mu} = -V^\nu K_{\mu;\nu} = -V^\nu \phi_{;\nu} V_\mu - \phi \dot{V}_\mu = -\phi \dot{V}_\mu$$

$$\phi_{;\mu} = -\phi \dot{V}_\mu$$

It then follows that the gradient of the inactive vector associated with time in the co-moving frame is:

$$K_{\mu;\nu} = \phi_{;\nu} V_\mu - \phi_{;\mu} V_\nu + \phi\omega_{\mu\nu} + \phi\theta_{\mu\nu} \,.$$

Since this is anti-symmetric, again I see that the symmetric compression coefficient contributions are zero.

I compute the second covariant derivative of the vectors using an identity that relates the curvature to the difference of second order derivatives with an interchange of order:

$$K_{\mu;\nu\lambda} - K_{\mu;\lambda\nu} = \phi R_{\mu\sigma\lambda\nu} V^\sigma \,.$$

These equations are then used together to deduce the equations for the gravitational field in terms of the sources (curvature and vorticity):

$$h^{\mu\nu}\phi^{-1}\phi_{;\mu\nu} = R_{\mu\nu} V^\mu V^\nu + \omega_{\mu\nu}\omega^{\mu\nu} \,.$$

The argument leading to this equation is general, depending only on the assumption that there is a Killing vector proportional to the flow.

However, in the single strategy model, this is the first equation of Eq. (D.4):

$$\frac{df}{du} = 2\kappa\left(\mu_0 + p_0 - \frac{\mu_0 - p_0}{D-2}\right) - 4\omega_\alpha\omega_\alpha \,.$$

This is a very surprising result for game theory, though on reflection, it is consistent with how a dynamic theory should work. There should be a force that moves non-rational behavior towards a rational fixed point. For the theory of games, a good name for this force might be the ***rational***

field. This is clearly a new distinction and potentially an important and powerful distinction. With the centrally co-moving hypothesis, the measure of the ***rational force*** is expressed in terms of the time component of the co-moving metric (the "co-moving gravitational field"):

$$f = \frac{1}{\gamma_{00}}\frac{d\gamma_{00}}{du} = \frac{2}{\phi}\frac{d\phi}{du}. \quad \text{(D.6)}$$

Related to the notion of the rational field is the transverse volume expansion, which is determined by the trace of the transverse market field:

$$\frac{d\Theta}{du} = -2\kappa(\mu_0 - p_0) + 4\omega_\alpha\omega_\alpha .$$

This allows the deduction that market density ($\mu_0 - p_0 \geq 0$), independent of the dimension of space–time, induces transverse contraction and ***vorticity*** induces transverse expansion. The more market density, the stronger the rational field and hence contraction occurs. In particular, it suggests that the equilibrium point is the point to which market density contracts. I caution however that there may be other effects that counter balance rational behavior.

D.3 Thermal Properties

I will show that the thermal equilibrium property Eq. (A.4) holds for a class of isentropic fluids:

$$\frac{1}{T}\frac{dT}{dp} = \frac{1}{\mu + p}. \quad \text{(D.7)}$$

Requiring thermal equilibrium is a constraint on the system. A system not in thermal equilibrium is a system in which there is heat flow from one part of the system to another. In the models considered here that are in thermal equilibrium, there is no heat flow but there is a temperature gradient. The gradient exists to prevent heat flow from one part of the system to another. This is not something we are familiar with. Scientists normally think of thermal equilibrium being defined as a state of a substance in which all parts have the same temperature. This is a consequence of thinking of heat as a form of energy that has no mass and hence creates no gravity. Astronomers might consider such effects but even there the effects are small. However for games, there is no guarantee that such effects will be small.

To create thermal equilibrium, I start with the general form of the single strategy model (with both vorticity and expansion parameters) and the form of the equilibrium condition Eq. (A.3).

$$\int\left(\frac{\mathbf{V}}{T}\delta\mu + \frac{\mu+p}{T}\delta\mathbf{V}\right) = 0 .$$

The form of the volume element in the single strategy model is:

$$\mathbf{V} = V^{\mu} d\sigma_{\mu}$$
$$d\sigma_k = \sqrt{g\gamma}\sum(-)^{sig} dx^1 \cdots du \cdots dx^j = \gamma\sum(-)^{sig} dx^1 \cdots du \cdots dx^j$$
$$d\bar{\sigma}_k \equiv \sum(-)^{sig} dx^1 \cdots du \cdots dx^j$$
$$d\sigma_k = \gamma d\bar{\sigma}_k$$
$$\mathbf{V} = \gamma V^k d\bar{\sigma}_k$$

The volume element is the projection of the flow vector onto a hypersurface in the multidimensional space of the game. The volume element depends on the determinant of the full metric, which simplifies in the harmonic gauge: $\sqrt{g\gamma} = \gamma$.

The analysis below is based on a specific assumption that the changes to the metric are made without changing the flow vector, $\delta V^j = 0$ and are orthogonal to the flow, $V^j \delta\gamma_{jk} = 0$. I change only the co-moving volume element. I imagine the volume at any point along the active strategy. At each point, I could look at the volume in the appropriate rest frame, in which case the volume is proportional to $\gamma e^{-\nu}$. The first factor is the determinant that makes the volume element invariant and the second factor is the length of the time-like vector in the transverse evolution frame. The scalar is determined by the connection:

$$f = C_{jk} V^j V^k = \frac{d\gamma_{jk}}{du} V^j V^k = 2\frac{d\nu}{du} . \tag{D.8}$$

This makes sense when I am in fact in the rest frame; the determination however is defined in all frames. The variation is zero by the assumption that the metric variations are orthogonal to the flow.

The variation of the volume element can now be performed and the resultant condition expressed in terms of a common volume element:

$$\int d\bar{\sigma}_k a^k \gamma \vartheta \left(\delta\mu + (\mu + p)\gamma^{-1}\delta\gamma \right) = 0$$
$$\vartheta \equiv \frac{1}{Te^{\nu}}$$

I now transform to reduced pressure and energy density, Eq. (D.2):

$$\int d\bar{\sigma}_k a^k \vartheta \left(\delta\mu_0 + p_0\gamma^{-1}\delta\gamma \right) = 0 .$$

The meaning of the reduced variable is clear. They are in fact the energy density and pressure of the problem where the volume element change is given by the change in the determinant $\delta \ln \gamma$.

The difficult part of the calculation is the computation of the energy density as defined by $\mu = T_{\mu\nu}V^{\mu}V^{\nu}$ in terms of the metric. This is done using the field equations:

$$\kappa\mu_0 = \kappa\gamma T_{\mu\nu}V^{\mu}V^{\nu} = -V^j V^k \gamma \hat{R}_{jk} + \tfrac{1}{2}\left(\gamma g_{ab} \hat{R}^{ab} + \gamma\gamma^{jk} \hat{R}_{jk} \right)$$

$$g_{ab}\gamma \hat{R}^{ab} = -\tfrac{1}{2}\gamma^{jk}\frac{d^2\gamma_{jk}}{du^2} + \tfrac{1}{4}\gamma^{-1}\frac{d\gamma}{du}\gamma^{-1}\frac{d\gamma}{du} - \tfrac{1}{4}\frac{d\gamma^{jk}}{du}\frac{d\gamma_{jk}}{du}$$

$$V^j V^k \gamma \hat{R}_{jk} = -\tfrac{1}{2} V^j V^k \gamma_{jm} \frac{dC^m_k}{du}$$

$$\gamma^{jk}\gamma \hat{R}_{jk} = -\tfrac{1}{2}\frac{dC^k_k}{du}$$

There are contributions from the active curvature $\hat{R}^{ab}$ and the inactive curvature $\hat{R}_{jk}$. The results can be combined to determine the energy density:

$$2\kappa\mu_0 = \begin{pmatrix} V^j V^k \dfrac{dC_{jk}}{du} - V^j V^k \gamma^{pq} C_{jp} C_{kq} \\ -\tfrac{1}{2}\gamma^{jk}\dfrac{dC_{jk}}{du} + \tfrac{1}{4}C^{jk}C_{jk} + \tfrac{1}{4}\dfrac{d\ln\gamma}{du}\dfrac{d\ln\gamma}{du} - \tfrac{1}{2}\dfrac{d^2 \ln\gamma}{du^2} \end{pmatrix} .$$

The exercise is to compute the metric variations $\delta\gamma_{jk}$ that are orthogonal to the flow $V^j\delta\gamma_{jk}=0$. Once complete, coefficients of the variations can be simplified using the solutions to the field equations for a perfect fluid. The form can be expressed in terms of the projection of the orthonormal set of vectors U^j_α introduced earlier [*Cf.* Eq. (D.4)]:

$$\int \vartheta d\bar{\sigma}_s\left(2\kappa\delta\mu_0+2\kappa p_0\delta\ln\gamma\right)=\int \vartheta d\bar{\sigma}_s E_{\alpha\beta}U^j_\alpha U^k_\beta\delta\gamma_{jk}=0.$$

The computation is lengthy. I quote the result:

$$E_{\alpha\beta}=\begin{pmatrix}\frac{1}{\vartheta}\frac{d^2\vartheta}{du^2}\delta_{\alpha\beta}+\frac{1}{2}\frac{1}{\vartheta}\frac{d\vartheta}{du}\frac{d\ln\gamma}{du}\delta_{\alpha\beta}\\ +4\omega_\alpha\omega_\beta+4\theta_\alpha\theta_\beta+\frac{1}{2}e_{\alpha\beta}\frac{1}{\vartheta}\frac{d\vartheta}{du}\end{pmatrix}=0.$$

This is a tensor relationship and has solutions in particular if the expansion and vorticity vanish:

$$\theta_\alpha=\omega_\alpha=0.$$

I express the resultant equation in terms of the shear and compression:

$$\frac{d^2\ln\vartheta}{du^2}+\left(\frac{d\ln\vartheta}{du}\right)^2+\frac{1}{2}\frac{d\ln\vartheta}{du}\left(\frac{D-3}{D-2}\Theta+f\right)=0$$
$$\frac{1}{2}\frac{d\ln\vartheta}{du}\Sigma_{\alpha\beta}=0$$

The solution allows for there to be a constant but non-zero transverse shear (this is because the vorticity vanishes) if the temperature satisfies:

$$\frac{d\ln\vartheta}{du}=0\Leftrightarrow\frac{dv}{du}+\frac{1}{T}\frac{dT}{du}=0\Leftrightarrow\frac{1}{T}\frac{dT}{du}=-\frac{1}{2}f=\frac{1}{\mu+p}\frac{dp}{du}.$$

I derive the form Eq. (A.4). This form, which was derived with vorticity and expansion excluded, can be extended to include them if the change in (reduced) entropy is equal to the variation of the scalar $\varphi_0\equiv\frac{1}{2\kappa}\left(df/du+4\omega_\alpha\omega_\alpha-4\theta_\alpha\theta_\alpha\right)$. I thus improve on Tolman's analysis to some degree since I make no assumption about the relationship between the energy density and pressure and provide a way to view the expansion and vorticity.

D.4 Extensions to Perfect Fluid

The solution of the Single Strategy Model in thermal equilibrium with no entropy flow specifies a relationship between the purely thermodynamic quantities of temperature, pressure and density, Eq. (D.7). This implies that the entropy and temperature together satisfy:

$$dU = TdS - pdV . \tag{D.9}$$

The theory as described includes the definition of internal energy, volume and temperature, thereby determining entropy, though a formula for the entropy is not actually determined nor needed to specify the energy density and pressure relationship. Classical thermodynamics starts from the conservation of energy and the assertion that the energy $U(S,V)$ in the above equation is a function of two independent variables, the entropy S and volume V. The temperature and pressure are dependent variables determined by differential relations in terms of the energy function:

$$\begin{aligned} T(S,V) &= \left(\frac{\partial U}{\partial S}\right)_V \\ p(S,V) &= -\left(\frac{\partial U}{\partial V}\right)_S \end{aligned} .$$

The differential relation for temperature is at constant volume and the differential relation for pressure is at constant entropy. Knowledge of the energy as a function of its independent variables determines all thermodynamic quantities.

In practice, it is often the case as it is here that information is known about certain relationships between the thermodynamic variables. For the single strategy model with no vorticity or expansion, Eq. (D.7) determines the variation of temperature with pressure at constant entropy:

$$\left(\frac{\partial T}{\partial p}\right)_S = \frac{T}{\mu + p} . \tag{D.10}$$

This shows that the temperature increases in response to an increase in pressure. The coefficient is related to the differentials of the energy:

$$\left(\frac{\partial T}{\partial p}\right)_S = \frac{\left(\frac{\partial T}{\partial V}\right)_S}{\left(\frac{\partial p}{\partial V}\right)_S} = -\frac{\left(\frac{\partial}{\partial V}\left(\frac{\partial U}{\partial S}\right)_V\right)_S}{\left(\frac{\partial}{\partial V}\left(\frac{\partial U}{\partial V}\right)_S\right)_S} = \frac{V\left(\frac{\partial U}{\partial S}\right)_V}{U-\left(\frac{\partial U}{\partial V}\right)_S V}.$$

Thus the single strategy model solution imposes a somewhat complex constraint on the energy function. The goal in classical thermodynamics is by means of such mathematical arguments to deduce the form of the energy function using a minimal number of inputs including constraints such as the one above.

I start with a model that illustrates that there are solutions for the energy which satisfy the above constraint. In this example I assume a constant ***market overhead index*** α_c:

$$\mu = \alpha_c p .$$

This defines the relationships between pressure, matter density and temperature:

$$\frac{\rho}{\rho_c} = \left(\frac{p}{p_c}\right)^{\frac{\alpha_c}{\alpha_c+1}}, \quad \frac{T}{T_c} = \left(\frac{p}{p_c}\right)^{\frac{1}{\alpha_c+1}}, \quad \frac{\rho}{\rho_c}\frac{T}{T_c} = \frac{p}{p_c}. \tag{D.11}$$

The values x_c with subscripts are constants. The first equation follows from the conservation of energy Eq. (D.9) and the statement that zero entropy change determines the matter density in terms of the pressure:

$$TdS = 0 = d\left(\frac{\mu}{\rho}\right) + pd\frac{1}{\rho}.$$

The second equation follows from Eq. (D.10), which relates the pressure and temperature. The last equation is a consequence of the first two, called the ***equation of state*** for a ***perfect fluid***. For numerical examples I will focus on such fluids as being simple possibilities that express a relationship between energy density and pressure in terms of a single parameter, the market overhead index.

The above argument is suggestive. To see more clearly the possibilities that exist for strategic-fluids, I extend the discussion of thermodynamics from Appendix A using results from classical thermodynamics [*e.g.* Pippard (1961)]. The classical treatment obtains results which are primarily of a mathematical nature[44] that follows from the form of the conservation of energy and the concept that such an equation defines the internal energy $U(S,V)$ as a function of two independent variables, the entropy S and $V=\rho^{-1}$ volume.

The mathematical machinery is based on the requirement that the internal energy be a given function of the stated variables, which imposes a condition that the second partial derivatives commute:

$$\left(\frac{\partial}{\partial V}\left(\frac{\partial U}{\partial S}\right)_V\right)_S = \left(\frac{\partial}{\partial S}\left(\frac{\partial U}{\partial V}\right)_S\right)_V \Leftrightarrow \left(\frac{\partial T}{\partial V}\right)_S = -\left(\frac{\partial p}{\partial S}\right)_V .$$

These are a subset of what are called the ***Maxwell thermodynamic relations***. The full set is obtained by making successive changes of variables to allow different pairs of variables to be independent. One change is accomplished by introducing the function (which in the literature is called a "Gibbs" free energy) $G(p,T)\equiv U+pV-TS$, which is a known function of the pressure and temperature whenever the energy function is given. Conversely given such a function of the pressure and temperature, the energy can be computed.

One starts with the statement that the "Gibbs" free energy determines the entropy and volume using the differential relations based on temperature and pressure being independent:

$$dG=-SdT+Vdp, \quad S=-\left(\frac{\partial G}{\partial T}\right)_p, \quad V=\left(\frac{\partial G}{\partial p}\right)_T .$$

Using these values, the energy density is computed as a function of the pressure and temperature scalars: $\mu(p,T)=V^{-1}(G-pV+TS)$. The "Gibbs" free energy is convenient whenever one knows information about various thermodynamic coefficients in terms of pressure and temperature.

---

[44] This is of course what we want here. We are not trying to use physics as an analogy but use it for its mathematical machinery.

Equivalently, the entropy and pressure are computed in terms of the free energy $F(V,T)=U-TS$, which is a function of temperature and volume:

$$dF=-SdT-pdV,\quad S=-\left(\frac{\partial F}{\partial T}\right)_V,\quad p=-\left(\frac{\partial F}{\partial V}\right)_T.$$

The latter equation is a good starting point if we have information about the equation of state: The pressure is given as a function of temperature and volume. The first equation is a good starting point if we focus on the relationship between volume and temperature for constant entropy.

I give an example of the power of the approach by working with the "Gibbs" free energy and assuming the independent variables are the pressure and temperature:

$$\left(\frac{\partial T}{\partial p}\right)_S=-\left(\frac{\partial T}{\partial S}\right)_p\left(\frac{\partial S}{\partial p}\right)_T=\left(\frac{\partial T}{\partial S}\right)_p\left(\frac{\partial V}{\partial T}\right)_p.$$

Standard textbooks [e.g. Pippard (1961)] allow me to write Eq. (D.10) in terms of coefficients which are determined using these variables.

This shows that the relationship set by change at constant entropy is set by the behaviors of the expansion coefficient $\beta=\left(\partial V/\partial T\right)_p/V$ and specific heat $c_p=T\left(\partial S/\partial T\right)_p$ at constant pressure. These concepts allow a systematic study of the possible relationships.

In this example, I look for a general solution that has a perfect fluid equation of state $pV=n_cT$ satisfying Eq. (D.10). The expansion coefficient $\beta=T^{-1}$ and the isothermal compressibility $k_T=-\left(\partial V/\partial p\right)_T/V$ are determined by the equation of state. These coefficients determine the difference between the specific heats at constant pressure and volume:

$$c_p-c_V=\frac{VT\beta^2}{k_T}=n_c.$$

The perfect fluid equation of state is a partial differential equation for the free energy:

$$p=-\left(\frac{\partial F}{\partial V}\right)_T=\frac{n_cT}{V}.$$

This can be solved to obtain the free energy in terms of an unknown function $\theta(T)$ of temperature:

$$F(V,T) = -n_c T \ln\frac{V}{V_c} + \theta(T).$$

The claim is that all of the thermodynamic quantities follow from this equation and should provide sufficient constraints to determine an equation for the unknown function.

First I note that the entropy is determined:

$$S(V,T) = n_c \ln\frac{V}{V_c} - \theta'(T).$$

From the entropy I determine the specific heats:

$$\begin{aligned} c_V &= -T\theta''(T) \\ c_p &= n_c - T\theta''(T) \end{aligned}.$$

I also determine the total energy:

$$U = F + TS = \theta(T) - T\theta'(T).$$

I have enough information to evaluate Eq. (D.10), which provides the fundamental equation for a perfect fluid consistent with the single strategy model:

$$\begin{aligned} &\left(\frac{\partial T}{\partial p}\right)_S = \frac{\beta VT}{c_p} = \frac{V}{n_c - T\theta''(T)} = \frac{VT}{U + pV} \\ &U + pV = n_c T - T^2\theta''(T) \\ &T^2\theta''(T) - T\theta'(T) + \theta(T) = 0 \end{aligned}.$$

The most general solution is in terms of two constants $\{\alpha_c,\ T_c\}$:

$$\theta = -n_c \alpha_c T \ln\frac{T}{T_c}.$$

The resultant specific heats are constant:

$$\begin{aligned} c_V &= T\theta''(T) = \alpha_c n_c \\ c_p &= n_c + \alpha_c n_c \end{aligned}.$$

The resultant energy density is proportional to pressure:

$$U = n_c \alpha_c T$$
$$\mu = \alpha_c p \quad .$$

I obtain the example described earlier in which the market overhead index is constant. I obtain more however. I determine that the specific heats are constant and their ratio is set by the overhead:

$$\gamma \equiv \frac{c_p}{c_V} = \frac{1+\alpha_c}{\alpha_c} = 1 + \alpha_c^{-1} .$$

I summarize the result for the free energy:

$$F = -n_c T \ln \frac{V}{V_c} - \alpha_c n_c T \ln \frac{T}{T_c} . \tag{D.12}$$

For later reference, I observe that a characteristic of this model is that larger values of the market overhead index correspond to larger specific heats. This gives an interpretation of the market overhead index. A large index indicates that an increase in temperature generates more heat than a correspondingly lower index. It provides some justification for the terminology, since it is natural to ascribe overhead to wasted effort, effort that does not generate movement of strategic-mass.

I believe it is important to have more general models in mind as well, models which are also consistent with the single strategy model. I obtain a general solution to the constraint Eq. (D.10). The natural independent variables for this constraint are the pressure and entropy, which requires the introduction of what is called the enthalpy $H(p,S) = U + pV$ in classical physics:

$$dH = TdS + Vdp \Rightarrow T = \left(\frac{\partial H}{\partial S}\right)_p , \quad V = \left(\frac{\partial H}{\partial p}\right)_S . \tag{D.13}$$

With some rearranging the single strategy solution constraint is:

$$\frac{T}{c_p}\left(\frac{\partial V}{\partial T}\right)_p = \frac{T}{\mu + p} \Rightarrow \frac{1}{c_p}\left(\frac{\partial V}{\partial T}\right)_p = \frac{V}{U + pV} = \frac{V}{H}$$
$$\Rightarrow H\left(\frac{\partial V}{\partial S}\right)_p = V\left(\frac{\partial H}{\partial S}\right)_p \quad .$$

The resultant equation is a second order partial differential equation:

$$HH_{pS} = H_p H_S . \tag{D.14}$$

The subscripts indicate partial derivative with the other independent variable held fixed. I obtain a solution by considering $\Phi = \ln H$:

$$\begin{aligned} &e^{\Phi}\left(e^{\Phi}\Phi_S\Phi_p + e^{\Phi}\Phi_{pS}\right) = \Phi_p\Phi_S e^{2\Phi} \\ &\Phi_{pS} = 0 \end{aligned}.$$

I integrate this twice to obtain the general solution:

$$\begin{aligned} &\Phi_p = \theta'(p)/\theta(p) \\ &\Phi = \ln\theta(p) + \ln\psi(S) = \ln H \end{aligned}.$$

The general solution is thus a product of two arbitrary functions:

$$\begin{aligned} &H = \theta(p)\psi(S) \\ &V = \theta'(p)\psi(S) \\ &T = \theta(p)\psi'(S) \end{aligned}. \tag{D.15}$$

The solution determines the energy density and in particular shows that it is independent of entropy:

$$\mu(p,S) = \frac{H - pV}{V} = \frac{\theta(p)}{\theta'(p)} - p. \tag{D.16}$$

The energy density and the corresponding market overhead index are functions of pressure only:

$$\alpha(p) = \frac{d\mu}{dp} = 1 - \frac{\theta\theta''}{\theta'^2} - 1 = -\frac{\theta\theta''}{\theta'^2}. \tag{D.17}$$

I verify that the equations that I have started with are both satisfied:

$$\begin{aligned} &\frac{1}{\rho}\left(\frac{\partial\rho}{\partial p}\right)_S = -\frac{1}{V}\left(\frac{\partial V}{\partial p}\right)_S = -\frac{\theta''}{\theta'} = \frac{-\dfrac{\theta\theta''}{\theta'^2}}{\dfrac{\theta(p)}{\theta'(p)}} = \frac{\alpha}{\mu + p} = \frac{1}{\mu + p}\left(\frac{\partial\mu}{\partial p}\right)_S \\ &T\left(\frac{\partial p}{\partial T}\right)_S = \frac{\theta\psi'}{\theta'\psi'} = \frac{\theta}{\theta'} = \mu + p \end{aligned}.$$

I record some of the thermodynamic coefficients for reference. The specific heat for constant pressure is a function only of entropy:

$$\frac{1}{T}\left(\frac{\partial T}{\partial S}\right)_p = \frac{\psi''(S)}{\psi'(S)} = \frac{1}{c_p} \Rightarrow c_p = \frac{\psi'}{\psi''}.$$

In contrast, the specific heat for constant volume is in general a function of both entropy and pressure:

$$c_V = \frac{\psi'}{\psi'' - \dfrac{\theta'^2 \psi'^2}{\theta \psi \theta''}}.$$

The ratio of specific heats $\gamma = c_p / c_V$ is:

$$\gamma = 1 - \frac{\theta'^2 \psi'^2}{\theta \psi \theta'' \psi''}.$$

The equation of state is the relationship between the volume, temperature and pressure and is expressed parametrically by knowledge of volume $V(p,S)$ and temperature $T(p,S)$ as a function of pressure and entropy and eliminating the entropy from these known functions.

There is one effect that might occur in the dynamic theory of games, which is related to the ***Joule–Thomson effect***[45]:

$$\left(\frac{\partial T}{\partial V}\right)_U = -\frac{1}{c_V}\left(T\left(\frac{\partial p}{\partial T}\right)_V - p\right) = p\frac{\psi''}{\psi'} + \frac{\theta'\psi'}{\theta''\psi}\left(1 - p\frac{\theta'}{\theta}\right). \quad \text{(D.18)}$$

The Joule effect describes the change in temperature when a gas expands into an empty area. It changes its volume irreversibly. It is related to the Joule–Thomson effect, which is the basis of refrigeration. In the Joule–Thomson experiment, a gas jet at high pressure goes through a porous plug or throttle to an area with a lower pressure. There is a temperature drop described by:

$$\left(\frac{\partial T}{\partial p}\right)_H = \frac{1}{c_p}\left(T\left(\frac{\partial V}{\partial T}\right)_p - V\right) = \theta'\left(\psi' - \frac{\psi\psi''}{\psi'}\right). \quad \text{(D.19)}$$

[45] For a classical discussion of this effect, see Landau-Lifshitz (1958) and under the name of the Joule-Kelvin effect and related effects see Pippard (1961).

I think that it is an interesting problem to translate these concepts into the theory of games: What would such an expansion mean? It might mean that under a rapid loss of strategic–density, the resources needed to produce a unit of production also drops. In any case, the effect is independent of the unknown function.

I give one class of models in which such effects can arise. I take the specific heat at constant pressure to be a constant c_p and the entropy factor to be determined by this constant:

$$\psi = \psi_c e^{S/c_p} .$$

From the general solution Eq. (D.15), I determine the equation of state in terms of the energy density:

$$(\mu + p)V = c_p T .$$

The ratio of specific heats is independent of the entropy for this model and is determined in terms of the market overhead:

$$\gamma = 1 - \frac{\theta'}{\theta''}\frac{\theta'}{\theta} = 1 + \frac{1}{\alpha(p)} .$$

The Joule effect, Eq. (D.18) is determined by the energy density and market overhead index:

$$\left(\frac{\partial T}{\partial V}\right)_U = p\frac{\psi''}{\psi'} + \frac{\theta'\psi'}{\theta''\psi}\left(1 - p\frac{\theta'}{\theta}\right) = \frac{1}{c_p}\left(\frac{\theta'}{\theta''} + \gamma p\right) = -\frac{1}{\alpha c_p}(\mu - \alpha p) .$$

Thus the sign of the effect is determined by the sign of the ***power defect*** $\mu - p d\mu/dp$. The model in which the energy density is a constant times the pressure is a dividing line between possibilities.

I think there might be some evidence in organizations for this effect. An ***effectiveness of power*** or ***limits to power*** principle is generally recognized and states that increased political power or control produces progressively less increase in strategic–density. If this is done for a constant number of resources per unit of production, then this reflects an isothermal change. The equation of state can be used to deduce that the energy density has a component that now decreases less fast at large pressures than at small.

I provide an example of this behavior:

$$\left(\frac{\alpha_c}{1+\beta\sqrt{p}}+1\right)p=cT\rho\,.$$

This is the equation of state, from which the energy density, the market overhead and the power defect follow:

$$\mu=\frac{\alpha_c p}{1+\beta\sqrt{p}},\quad \alpha=\alpha_c\frac{1+\frac{1}{2}\beta\sqrt{p}}{\left(1+\beta\sqrt{p}\right)^2},\quad \mu-\alpha p=\alpha_c\,p\frac{\frac{1}{2}\beta\sqrt{p}}{\left(1+\beta\sqrt{p}\right)^2}\,.$$

For large pressures the market overhead index approaches zero, reflecting the increased control:

$$\alpha\to\frac{\frac{1}{2}\alpha_c}{\beta\sqrt{p}}\,.$$

In the same limit, the Joule effect increases linearly as a consequence, showing that a large decrease in strategic–density (increase in volume) results in the number of resources per unit of production (temperature) dropping:

$$\left(\frac{\partial T}{\partial V}\right)_U=-\frac{1}{\alpha c}(\mu-\alpha p)\to-\frac{p}{c}\,.$$

At low pressures however, the effect is small. The market overhead approaches the constant α_c. In the same limit, the Joule effect goes to zero:

$$\left(\frac{\partial T}{\partial V}\right)_U=-\frac{1}{\alpha c}(\mu-\alpha p)\to-\frac{\frac{1}{2}\beta}{c}p^{\frac{3}{2}}\,.$$

The model provides insight into behaviors that are other than the assumed statement about the effectiveness of power.

Appendix E

Single Strategy Numerical Solutions

In this section I consider a game in thermal equilibrium. The example gives insight into behaviors that are consequences of being solutions to the full set of field equations. As a starting point, I choose a ***homogeneous model*** for the medium where the energy density μ is a function of pressure p. For any given function, I use the defining equations Eq. (A.6) of the fluid to compute the strategic–density ρ:

$$\frac{1}{\rho}\frac{d\rho}{du}=\frac{1}{\mu+p}\frac{d\mu}{du}. \tag{E.1}$$

If I associate the force that moves games towards equilibrium with rationality, then the strategic–density ρ is the source of this rationality; I think of strategic–density as a measure of experience, maturity or possibly even skill of the players. Any such measure generates "rest" energy or mass that is subject to gravitational force.

Furthermore, I assume that the market fluid is in thermal equilibrium, so that the system satisfies both the centrally co-moving hypothesis $\theta_\alpha = 0$ and the requirement of no vorticity $\omega_\alpha = 0$. In the example below, I take a constant market index of compressibility $\mu = \alpha p$. For the numerical calculation I choose $\alpha = 5$ as representative. If the index satisfies the constraint, $\alpha \geq 1$, I insure that the energy density will be larger than the pressure and will determine a negative transverse volume expansion based on the equation for the transverse volume expansion:

$$\frac{d\Theta}{du}=-(\alpha-1)\,p_0\,.$$

In other words, as I go further out along the radial deviation strategy, I get more contraction.

E.1 Fair Game Initial Conditions

Equations (D.5) for the fluid described above, can be simplified (in units $2\kappa = 1$):

$$\mu_0 = \alpha p_0$$
$$\omega_\alpha = \theta_\alpha = 0$$
$$\frac{d\Sigma_{\alpha\beta}}{du} = 0$$
$$\frac{d\Theta}{du} = -(\alpha - 1) p_0$$
$$\frac{df}{du} = \left(\alpha + 1 - \frac{\alpha - 1}{D - 2}\right) p_0$$
$$\frac{d \ln \gamma}{du} = \Theta + f$$
$$\frac{1}{T}\frac{dT}{du} = -\tfrac{1}{2} f$$
$$p_0 = \tfrac{1}{2} f\Theta + \tfrac{1}{4}\frac{D-3}{D-2}\Theta^2 - \tfrac{1}{4} Tr\mathbf{\Sigma}^2$$

I see that the transverse shear is constant and there are three coupled equations in $\{\Theta \quad f \quad \gamma\}$ to solve, given values for the market overhead index, the transverse shear components and initial conditions on $\{\Theta \quad f \quad \gamma\}$. The temperature is determined from this solution. The initial conditions are chosen so that the rational force is zero at the equilibrium position $u = 0$: $f(0) = 0$. I take two players each with two strategies and one value–choice, so the total number of dimensions is seven: $D = 7$. The number of independent metric components is 21 and the number of independent connection components is 21. There are five independent flow components, making the total number of unknowns and equations equal to 47.

The initial metric conditions are chosen to correspond to a fair game, whose game aspects I treat in Chapter 6. I define the strategic decomposition in Chapter 5. This provides a definition of the player strategic lengths and the radial and angular deviation strategies.

For a two-person zero-sum game in which the players each have two strategies, I choose the coordinates in the equilibrium frame to be $\{r_1 \;\; r_2 \;\; z_1 \;\; z_2 \;\; t\}$ from Chapter 5. For a fair game this can be taken to be the central frame. I rotate the frame to polar coordinates for the polar deviation components: This is the ***equilibrium polar frame*** which remains stationary under the transformation. I furthermore assume that the equilibrium motion is centered at an equilibrium radius r_0, where u, the active strategy, measures the deviation.

For the numerical model, I start with the initial conditions based on a representative fair game with initial coordinates and a unit vector flow in the player frame defined by:

$$\begin{gathered}\mathbf{x} = \{\tfrac{4}{3} \;\; -\tfrac{1}{3} \;\; -\tfrac{1}{3} \;\; \tfrac{4}{3} \;\; 0\} \\ \mathbf{V} = \sqrt{\tfrac{5}{22}}\{\tfrac{1}{3} \;\; \tfrac{2}{3} \;\; \tfrac{2}{3} \;\; \tfrac{1}{3} \;\; 8\}\end{gathered}.$$

In the equilibrium polar frame, the transformed coordinate and flows are:

$$\begin{gathered}\bar{\bar{\bar{\mathbf{x}}}} = \{1 \;\; 1 \;\; \sqrt{2} \;\; 0 \;\; 0\} \\ \bar{\bar{\bar{\mathbf{V}}}} = \sqrt{\tfrac{5}{22}}\{1 \;\; 1 \;\; 0 \;\; \tfrac{1}{10} \;\; 8\}\end{gathered}.$$

I have tweaked the result by adding a small angular flow, which of course requires renormalizing the flow, which is not done above so as to make apparent the origin of the flow values.

I now work in the inactive geometry space, with unit flow vector proportional to:

$$V^j \propto \left\{\frac{t \;\; \xi^1 \;\; \xi^2 \;\; r_1 \;\; r_2 \;\; \theta_\omega}{8 \;\; \tfrac{1}{10} \;\; \tfrac{1}{10} \;\; 1 \;\; 1 \;\; \tfrac{1}{10}}\right\}.$$

The value–choices are represented by $\{\xi^1 \;\; \xi^2\}$ and form the ***player subspace*** in the inactive geometry. The inactive metric is set in terms of the equilibrium distance $r_0 = \sqrt{2}$, the initial value of the gravitational potential $v_0 = \tfrac{1}{2}\ln \tfrac{7}{10}$ and the initial value of the determinant of the inactive metric γ_0.

I obtain the following form for the metric:

$\hat{\gamma}_{\mu\nu}$	t	ξ^1	ξ^2	r_1	r_2	θ_ω	u
t	e^{2v_0}	0	0	0	0	0	0
ξ^1	0	$-\lvert\gamma_0\rvert r_0^{-1} e^{-v_0}$	0	0	0	0	0
ξ^2	0	0	$-\lvert\gamma_0\rvert r_0^{-1} e^{-v_0}$	0	0	0	0
r_1	0	0	0	-1	0	0	0
r_2	0	0	0	0	-1	0	0
θ_ω	0	0	0	0	0	$-r_0^2$	0
u	0	0	0	0	0	0	$-\lvert\gamma_0\rvert$

The metric for the full model consists of both an active geometry and an inactive geometry. The active metric is the determinant of the inactive geometry

$$g^{uu} = -|\gamma| .$$

The mixed active/inactive components are set by the gauge potentials which are set to zero:

$$A_u^k = \left\{ \frac{t \quad \xi^1 \quad \xi^2 \quad r_1 \quad r_2 \quad \theta_\omega}{0 \quad 0 \quad 0 \quad 0 \quad 0 \quad 0} \right\}.$$

This leaves the inactive geometry components γ_{jk}, which are set by those components that involve the value–choices, those that don't and those which are a mixture. For this first model I take the metric components defined in terms of the inactive metric components $\bar{\gamma}_{\bar{j}\bar{k}} \equiv \gamma_{\bar{j}\bar{k}}$ defined on the subspace $\{\xi^1 \quad \xi^2\}$ to be diagonal and the diagonal components equal. The mixture is set by the market potentials in the equilibrium polar frame and is defined in terms of the inactive metric components $\bar{\gamma}_{\bar{j}\bar{k}}$ projected onto the mixed components $\gamma_{\bar{k}m}$:

$$\bar{A}_m^{\bar{j}} \equiv \bar{\gamma}^{\bar{j}\bar{k}} \gamma_{\bar{k}m} .$$

I take the boundary conditions for the vector potentials to be zero, $\bar{A}_m^{\bar{j}}(0) = 0$:

$$\gamma_{\bar{k}m}(0) = \bar{\gamma}_{\bar{k}\bar{j}} \bar{A}_m^{\bar{j}}(0) = 0 .$$

There is no loss in generality because the vector potential value at equilibrium does not enter into the market field.

I now have the sub-metric that depends on the strategic lengths, the polar angle and time. I assume for simplicity that in the polar equilibrium frame this subcomponent is diagonal and has the property that it is generated from the equilibrium frame by a rotation. This justifies the diagonal terms for the strategic lengths and the polar angle. The time component is assumed to be set by a potential well. Finally I set the value–choice components by the initial requirement that the inactive metric determinant be unity.

I assume the vector potentials are functions of the difference of the radial deviation strategy and the equilibrium deviation strategy distance $u = r_\omega - r_0$. The player market fields are determined in terms of the market potentials:

$$\bar{F}^{\bar{k}}_{um} = \frac{d\bar{A}^{\bar{k}}_{m}}{du} = \frac{d}{du}\left(\bar{\gamma}^{\bar{k}\bar{j}} \gamma_{\bar{j}m}\right) = \bar{\gamma}^{\bar{k}\bar{j}} C_{\bar{j}m} - \bar{\gamma}^{\bar{k}\bar{j}} C_{\bar{j}\bar{l}} \bar{A}^{\bar{l}}_{m} .$$

Thus the initial conditions for the derivative of the metric are determined by Eq. (C.2), the initial values of the player market fields:

$$C_{\bar{j}m}(0) = \bar{\gamma}_{\bar{j}\bar{k}} \bar{F}^{\bar{k}}_{um} .$$

In this example I presume that each player sees the same initial conditions and plays a ***realistic game***:

$$\begin{aligned} F^{\bar{k}}_{ur_1} &= 0 \\ F^{\bar{k}}_{ur_2} &= 0 \\ F^{\bar{k}}_{ut} &= 0 \\ F^{\bar{k}}_{u\theta_\omega} &= -r_0\omega(r_0) \equiv -r_0\omega_0 \\ r_0 &\equiv r_\omega(0) \end{aligned} .$$

For the specific models assumed here, I have both dynamic and thermal stability and hence no vorticity or expansion coefficients.

At equilibrium, I choose $f(0)=0$. These choices constrain the values of the connection components C_{jk}. I make an initial guess that the following solution determines a positive pressure[46]:

C_{jk}	t	ξ^1	ξ^2	r_1	r_2	θ_ω
t	0	0	0	0	0	0
ξ^1	0	0	0	0	0	$\omega_0 e^{-v_0}$
ξ^2	0	0	0	0	0	$\omega_0 e^{-v_0}$
r_1	0	0	0	$x+\omega_0 e^{-v_0}$	0	0
r_2	0	0	0	0	$x+\omega_0 e^{-v_0}$	0
θ_ω	0	$\omega_0 e^{-v_0}$	$\omega_0 e^{-v_0}$	0	0	0

I note the value $\omega_0 e^{-v_0} = 9\sqrt{10/7} \cong 10.76$. I have not however taken into account the constraints based on no vorticity or expansion parameters, nor have I specified the parameter x. This parameter is fixed by the initial value of pressure, which I take to be: $p_0 = 100$.

I take these constraints into account by working in the transverse evolution frame $\{V^j \quad U^j_\alpha\}$ where the connection is given by the transverse shear, transverse compression, vorticity and gravity force f. As noted, I set the vorticity and gravity force to be zero at equilibrium. The remaining shear and compression parameters are constrained by the market field. I find that the following co-moving connection components recreate the boundary conditions for the market field:

$c_{\alpha\beta}$	ξ^1	ξ^2	r_1	r_2	θ_ω
ξ^1	−10.75	0	0.03	0.03	8.27
ξ^2	0	−10.75	0.03	0.03	8.27
r_1	0.03	0.03	47.65	0	0
r_2	0.03	0.03	0	47.65	0
θ_ω	8.27	8.27	0	0	0

[46] For the presumed form, I require $|C_{jj}| \geq |\omega_0 e^{-v_0}|$ *for* $j = \{r_1, r_2\}$ for the strategic length components (if they have the same sign) in order that the initial pressure be non-negative. Otherwise these components are unconstrained.

The resultant connection reflects the twist that occurs when going from the transverse evolution frame back to the polar equilibrium frame by means of the orthonormal set $\{V^j \quad U^j_\alpha\}$:

C_{jk}	t	ξ^1	ξ^2	r_1	r_2	θ_ω
t	1.56	0	−0.0	−6.24	−6.24	−0.39
ξ^1	0	−9.09	−0.0	0	0	10.76
ξ^1	−0.0	−0.0	−9.09	0	0	10.76.
r_1	−6.24	0	0	48.76	1.11	0.28
r_2	−6.24	0	0	1.11	48.76	0.28
θ_ω	−0.39	10.76	10.76	0.28	0.28	0.02

I recover the desired result. The resultant connection matrix insures the initial condition on the market field, including no internal factions: The only non-zero field is the angular component. It also insures that the pressure is initially $p_0(0)=100$. The initial transverse compression that results is $\Theta(0)\cong -73.80$. The constant transverse shear matrix is:

$\Sigma_{\alpha\beta}$	ξ^1	ξ^2	r_1	r_2	θ_ω
ξ^1	25.51	0	−0.03	−0.03	−8.27
ξ^2	0	25.51	−0.03	−0.03	−8.27
r_1	−0.03	−0.03	−32.89	0	0
r_2	−0.03	−0.03	0	−32.89	0
θ_ω	−8.27	−8.27	0	0	14.76

The main transverse shear forces are the diagonal components and the large off-diagonal component associated with the value–choice and the polar angle. This matrix is constant along the active radial deviation strategy.

E.2 Fair Game Single Strategy Model

Solving the equations numerically leads to the following results. I define the radial deviation strategy in terms of the natural length u as follows: $d\ln r = r_0\sqrt{\gamma}du$. This leads to a new line element where the active contribution is $-dr^2/r^2$ consistent with the radius being a polar coordinate. This is a generalization of my initial approximation that $r \cong r_0 + u$. Since normally I am very close to the equilibrium position, I use the two terms interchangeably. For reference I show the radial deviation difference in terms of the natural length[47].

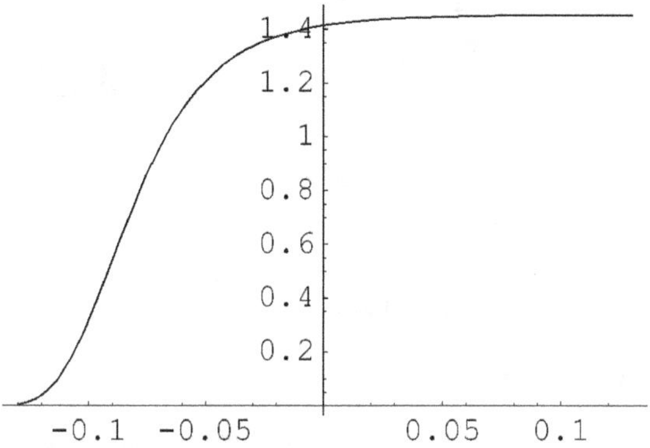

Fig. E.1 Radial Deviation Strategy (vertical axis) is shown versus natural length (horizontal axis).

Figure E.1 shows that the radial deviation strategy has a maximum a little larger than the equilibrium radial deviation strategy $r_0 = \sqrt{2}$. There is an interesting phenomenon that occurs at zero radial deviation strategy. A useful way to think of the variable u is that it takes on the values $(-\infty, \infty)$ and so parameterizes the distance in such a way that the most negative values are at the shortest "physical" distances and the most positive values are very far away. In actual calculations, there can be a minimum and maximum physical distance. I thus avoid saying at zero distance and at infinite distance. This says that the origin $u = 0$ is not at the closest possible distance but at the defined equilibrium. I show that the market substance is not collected at the physical origin but is circulating around the physical origin a short distance away.

[47] I use *Mathematica*® by Wolfram (1992) to compute the results.

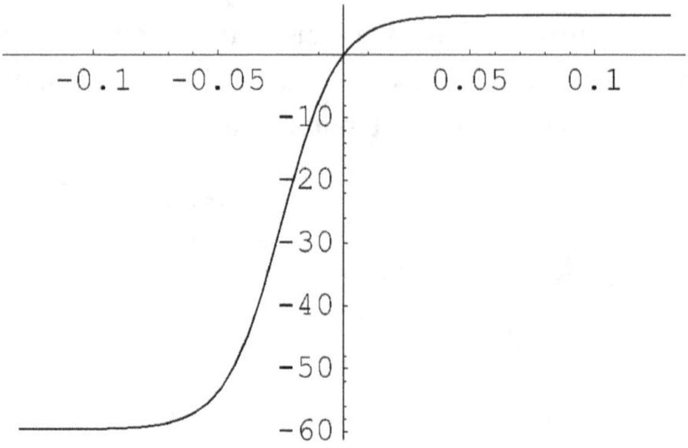

Fig. E.2 Rational force f (vertical axis) is shown versus natural length (horizontal axis).

Figure E.2 shows that the rational force is zero where the natural length is zero. There is a restoring force above and below the equilibrium radius. For this example, the restoring force below is much larger than the one above. The two forces become more equal as the initial pressure is increased.

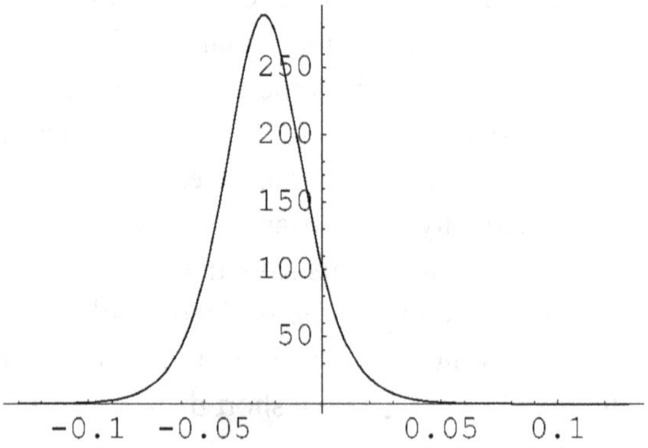

Fig. E.3 Reduced pressure p_0 (vertical axis) is shown versus natural length (horizontal axis).

Figure E.3 shows that the reduced pressure is maximal near the equilibrium natural length. The result is striking in that it predicts that the strategic–density is concentrated in a ring around $r_\omega = r_0$. This is seen clearly in Fig. E.4. The fact that the density of strategic-mass drops off sharply at negative values of natural length supports the proposed relationship between the radial deviation strategy and the natural length.

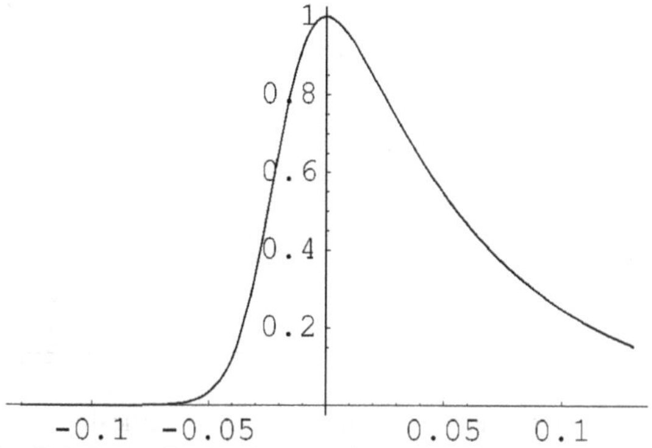

Fig. E.4 Strategic–density (vertical axis) is shown versus natural length (horizontal axis).

In the single strategy models, I have argued that temperature has a well defined meaning.

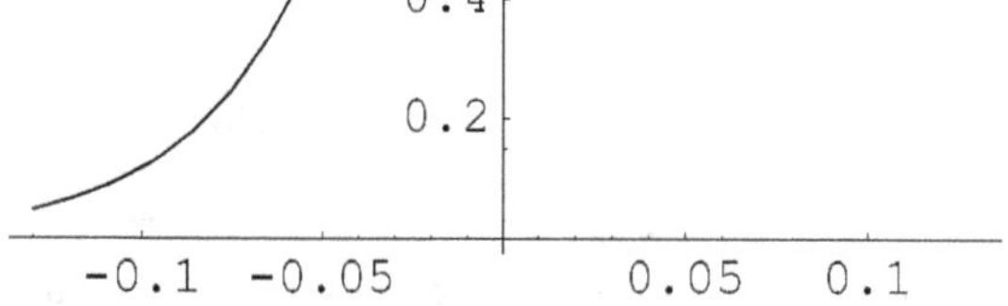

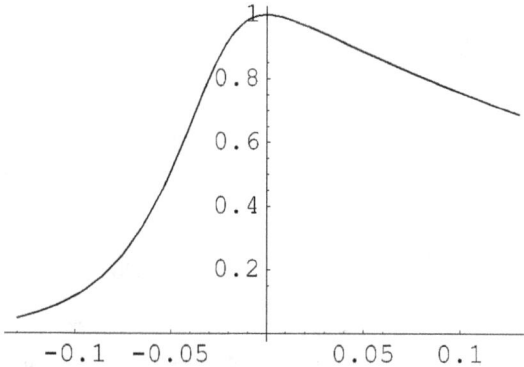

Fig. E.5 Temperature (vertical axis) is shown versus natural length (horizontal axis).

Figure E.5 shows there is a temperature gradient even at thermal equilibrium. The temperature is a maximum at the game equilibrium!

The transverse volume expansion parameter (Fig. E.6) reflects the fact that matter is collected around the equilibrium.

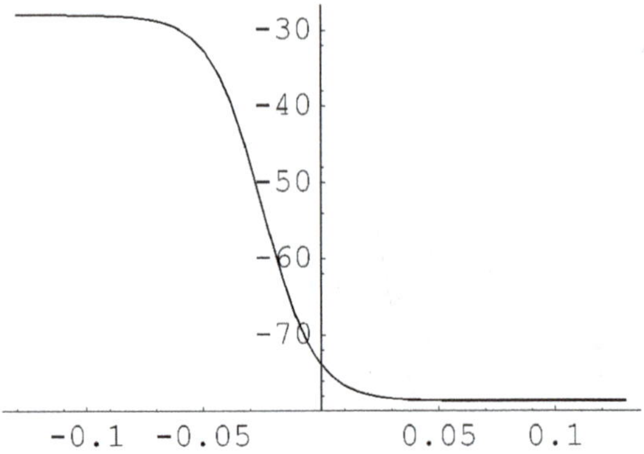

Fig. E.6 Transverse volume expansion (vertical axis) is shown versus natural length (horizontal axis).

It leads to the volume measure $\gamma = \det \gamma_{jk}$ decreasing to zero as the radial deviation increases.

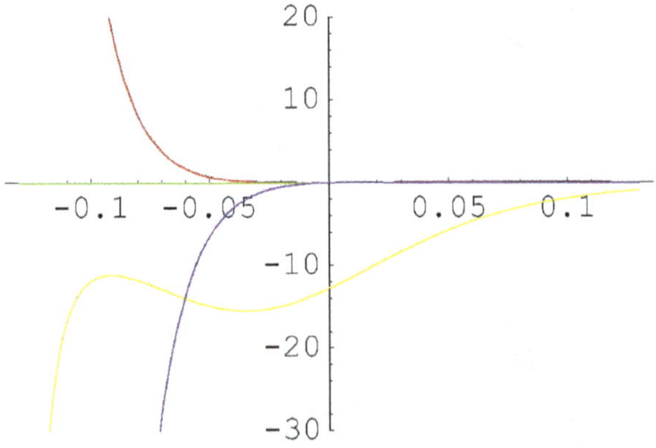

Fig. E.7 Player 1 game matrix elements (vertical axis) in the polar frame are plotted against the natural scale (horizontal axis)—Red, green, yellow and purple correspond to the market field $\overline{F}^{\bar{k}}_{um}$ along $\{r_1 = r_2 \quad r_\omega \quad \theta_\omega \quad t\}$ respectively.

The game matrix components are not constant as a function of the natural length (Fig. E.7).

The game matrix elements for player 2 are indistinguishable. For the exact solution, for all distances there is no field associated with the radial deviation. At equilibrium, the only non-zero field is the one associated with the polar angle.

The gravity field (Fig. E.8) determined by the gravitational force shows a "gravitational well" as a function of the natural length. The well is shallow based on the initial pressure chosen. It becomes more pronounced as the initial pressure is increased. This provides insight into how "market pressure" might be perceived in economic behavior.

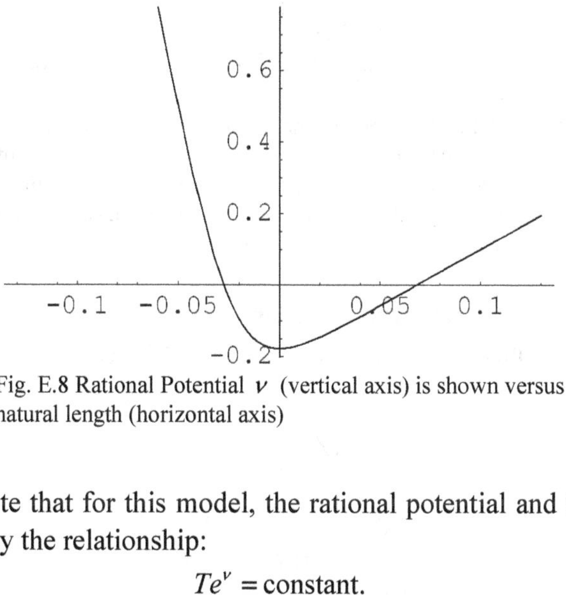

Fig. E.8 Rational Potential ν (vertical axis) is shown versus natural length (horizontal axis)

I also note that for this model, the rational potential and temperature are related by the relationship:

$$Te^{\nu} = \text{constant}.$$

This relationship explains why for thermal equilibrium, there is a non-trivial temperature gradient. The analogous effect in physics is apparently small except when dealing with truly large objects such as stars, galaxies, *etc.* but may be something of interest in an economic theory.

The flow vectors (Fig. E.9) are not at rest and so have a behavior that tracks with the temperature since they share the same equations with different boundary conditions. The flow is constant along a streamline.

The strategic length flows and the polar angular flows are quite small. This follows from the initial conditions.

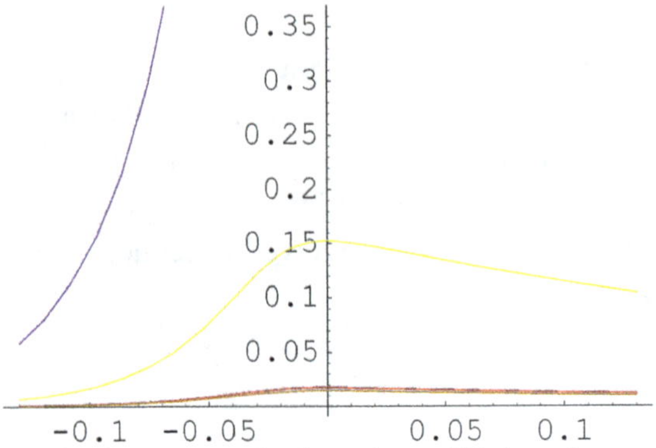

Fig. E.9 Flow vectors in the polar equilibrium frame are shown versus natural length)—purple, yellow on top of green, red on top of blue and brown are respectively, time, strategic lengths for player one and two, value–choice for player one and two and the polar angular flow.

The flows are relatively constant for distances larger than the equilibrium radius; the flows do drop to zero at large distance.

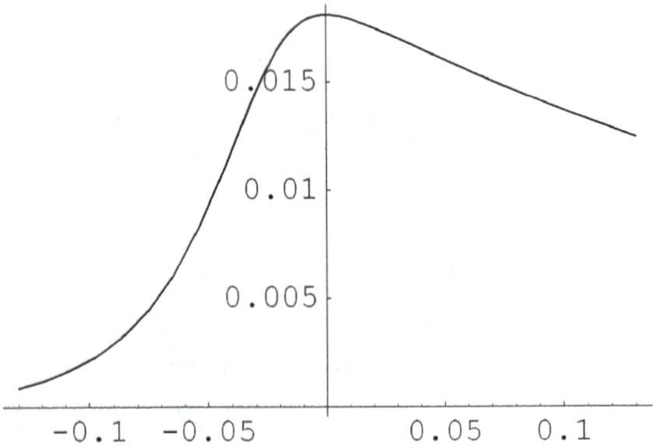

Fig. E.10 Player 1 "self-interest" (identical for player 2) is shown versus natural length.

I note in particular that the player self interest is maximal and represented by the player flows along the value–choice (Fig. E.10).

Appendix F

Streamlines

A general solution of the field equations $\hat{R}_{\mu\nu} - \frac{1}{2}\hat{\gamma}_{\mu\nu}\hat{R} = -\kappa T_{\mu\nu}$ exists subject to suitable gauge conditions on the metric and specification of boundary conditions on a suitable hypersurface. The field equations are supplemented with the equations for the flow, energy density and pressure, which follow from a model for the energy momentum tensor $T_{\mu\nu}$ and the imposition of the conservation laws $T^{\nu}_{\mu;\nu} = 0$. For a given metric, the conservation laws provide insight into the behavior of the solutions. I recall the flow equations for a perfect fluid, Eq. (3.8):

$$\begin{aligned} \dot{V}^{\mu} &= \frac{1}{\mu + p} h^{\mu\nu}\partial_{\nu}p \\ V^{\mu}\partial_{\mu}\mu + (\mu + p)V^{\mu}{}_{;\mu} &= 0 \end{aligned} .$$

I use the following definitions and normalizations:

$$\begin{aligned} \dot{V}^{\mu} &\equiv \frac{DV^{\mu}}{\partial s} \equiv V^{\mu}{}_{;\nu}V^{\nu} \\ h^{\mu\nu} &= \hat{\gamma}^{\mu\nu} - V^{\mu}V^{\nu} \\ \hat{\gamma}_{\mu\nu}V^{\mu}V^{\nu} &= 1 \end{aligned} .$$

If there are D dimensions in this space, the two conservation equations represent D independent first order partial differential equations in $D+1$ variables that consist of the $D-1$ independent flow variables and the two scalar variables pressure and energy density. There is one free function that can be specified at will.

Assuming that specification of the free function is made in an appropriate manner and the values of the unknown functions are specified on a hypersurface, the partial differential equations have a solution [*Cf.* Sommerfeld, (1964)]. In this appendix, the solution is shown to be given by coupled linear ordinary differential equations, which are amenable to standard techniques of numerical solution. This provides an important tool for obtaining insight into the behavior of these equations for dynamic games.

One gains insight into the solution by considering[48] ***streamlines***. I make the assumption that the unknown free function is specified. One example is that the energy density is constant. Another example is that the energy density is a known function of the pressure (such as $\mu = \alpha p$). I assume there is a D dimensional hypersurface on which the initial values of the coordinate, flows and remaining scalar function are specified. I furthermore assume the metric is specified everywhere in space. The ***streamlines*** are defined by the integral curves of the flow:

$$\frac{dx^{\mu}}{ds} = V^{\mu}.$$

By considering neighboring streamlines, a coupled set of linear ordinary differential equations is obtained for the flow gradients. In this way the streamline and all the unknown functions are determined.

F.1 General Treatment

In the treatment below, I consider the flow V^{μ} to be time-like. Initially I make no assumption about the particular form of the energy momentum tensor. In particular the discussion will be valid for the perfect fluid. The discussion is also valid more generally. As an example, it would apply to the case of a single inactive strategy (for a zero-sum game using a value–choice common to all players):

$$\dot{W}_a = \bar{h}_a^{\ b} \partial_b \phi + c_{\xi^0} e^{-\phi} \bar{F}_{ab} W^b; \quad \left(1 + \frac{\alpha - 1}{1 - \sigma^{-2} c_{\xi}^{\ 2} e^{-2\phi}}\right) \frac{d\phi}{d\tau} + W^a_{\ ;a} = 0. \text{(F.1)}$$

---

[48] There is a large literature on the subject of streamlines. There are a variety of notations and approaches, which go back into the history of the study of hydrodynamics. I generally follow the treatment of Hawking and Ellis (1973), though I differ from them in choice of metric convention. I have found Feynman (1963) helpful for providing a physical insight into streamlines.

For a single inactive strategy, the derived flow W^a is proportional to the active components V^a.

Hawking and Ellis (1973) provide a relativistic treatment of streamlines. Their application is to relativistic astrophysics, where the results are commonly known. The leap we are making here is that the mathematics leads to results that will have application to the dynamic theory of games. For this we take advantage of the fact that the results are based on geometric notions and are independent of the number of dimensions. I don't take this as self-evident however; for this reason I go through the arguments keeping in mind the goal to apply this to game theory. The essential idea is to find an orthonormal set of coordinate vectors that move along the streamline in the space of strategy–choices and time–choice. Given such a ***co-moving frame***, the goal is to describe the behavior of the flow gradients in this coordinate system.

I start with the timelike flow V^μ. This timelike flow describes an integral curve called a ***streamline***. Two streamlines are neighbors if they are related to each other by a "translation" specified by a commuting vector field Z^μ.

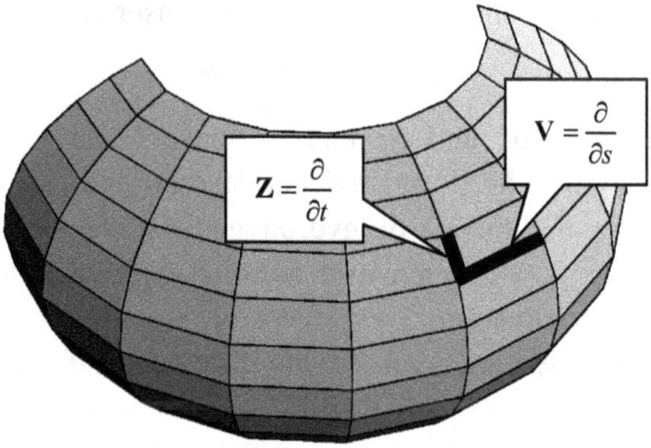

Fig. F.1 The grid provides a coordinate system.

The grid provides a coordinate system when the derivatives commute. This is the same as requiring that a small motion along one streamline, followed by a translation or jump to the other streamline is the same as doing the jump to the second streamline first followed by a small motion along that streamline:

$$Z^{\mu}{}_{;\nu}V^{\nu} = V^{\mu}{}_{;\nu}Z^{\nu}.$$

The variation of the translation vector along the streamline is determined by this covariant expression of the commutativity condition: $\dot{Z}_{\mu} = V_{\mu;\nu}Z^{\nu}$. I see that the variation of the translation along the flow is determined by the gradients of the flow. Two translations are equivalent if their transverse components achieve the same result.

In the discussion that follows, the flow vector is normalized to unity:

$$\hat{\gamma}_{\mu\nu}V^{\mu}V^{\nu} = 1.$$

The evolution of the transverse components

$${}_{\perp}Z^{\mu} = h^{\mu}_{\nu}Z^{\nu} = Z^{\mu} - V^{\mu}V_{\nu}Z^{\nu}$$

can be determined using this constraint:

$$\perp\left(\frac{D_{\perp}Z^{\mu}}{\partial s}\right) = \left(V^{\mu}{}_{;\nu} - \dot{V}^{\mu}V_{\nu}\right)_{\perp}Z^{\nu}. \qquad \text{(F.2)}$$

The gradient components of the flow are typically decomposed into a symmetric tensor $\theta_{\mu\nu}$ and an anti-symmetric tensor $\omega_{\mu\nu}$:

$$V_{\mu;\nu} - \dot{V}_{\mu}V_{\nu} \equiv \theta_{\mu\nu} + \omega_{\mu\nu}. \qquad \text{(F.3)}$$

The symmetric and anti-symmetric tensors have the property $\theta_{\mu\nu}V^{\nu} = \omega_{\mu\nu}V^{\nu} = 0$. The flow gradients describe the rate of "***strain***" of the fluid and in the theory of hydrodynamics, the decomposition is described as one into a symmetric ***volume expansion tensor*** and an antisymmetric ***vorticity tensor***.

The second order gradients of the flow depend on the order of differentiation: $V^{\mu}{}_{;\nu\rho} - V^{\mu}{}_{;\rho\nu} = R^{\mu}{}_{\sigma\rho\nu}V^{\sigma}$. The curvature of space–time dictates the size of the difference. I evaluate the second order deviation of the translation along the streamline and verify the result from the literature for the second derivative of the distance between flow lines:

$$h^{\mu}_{\nu}\frac{D}{\partial s}\left(h^{\nu}_{\rho}\frac{D}{\partial s}{}_{\perp}Z^{\rho}\right) = h^{\mu}_{\nu}\left(R^{\nu}{}_{\lambda\sigma\rho}V^{\lambda}V^{\sigma} + \dot{V}^{\nu}{}_{;\rho} - \dot{V}^{\nu}\dot{V}_{\rho}\right)_{\perp}Z^{\rho}. \qquad \text{(F.4)}$$

This provides a second order equation for the translation of one streamline into a neighbor. It allows the determination of the translation as a function of position along the streamline assuming information is known about the flows on the right hand side. The right hand side is determined in terms of the metric, the curvature and the conservation equations for the energy–momentum tensor which are known. The only statement about the dimensionality is implicit in the summation convention which goes over all the dimensions of space–time. Moreover, the curvature may have unique properties because of the dimensionality of the geometry.

F.2 Fermi Derivative

The above equations have a more transparent form in a frame with the following properties:

- One coordinate is along the direction of motion, that is, along the streamline.
- All other coordinates are transverse to the streamline.
- A set of vectors that start orthonormal at a point should stay orthonormal along the streamline.
- The equations should generalize those of a geodesic: For example, the metric should not change along the path.

Hawking and Ellis (1973) provide a formulation of the above equations that rests on the introduction of the ***Fermi derivative***, defined in terms of the covariant derivative and the flow vector:

$$\begin{gathered}\frac{D_F\mathbf{X}}{\partial s} \equiv \frac{D\mathbf{X}}{\partial s} + \hat{\gamma}\left(\mathbf{X}, \frac{D\mathbf{V}}{\partial s}\right)\mathbf{V} - \hat{\gamma}(\mathbf{X},\mathbf{V})\frac{D\mathbf{V}}{\partial s}\,. \\ \hat{\gamma}(\mathbf{X},\mathbf{V}) \equiv \hat{\gamma}_{\mu\nu}X^{\mu}V^{\nu}\end{gathered} \qquad \text{(F.5)}$$

The idea is that this derivative will allow the introduction of a fixed orthonormal coordinate system defined along the streamline. The vectors of that coordinate system will be constant under this derivative. I refer to this reference for the specific properties of this derivative. For our purposes it is sufficient to use the results.

The Fermi derivative allows an identification of a set of orthonormal vectors $\{\mathbf{E}_\alpha\}$ at an initial point on the streamline and insures that these vectors remain orthonormal along the streamline. I take one of these vectors to be the flow $\mathbf{E}_s = \mathbf{V}$, whose Fermi derivative can be shown to vanish. As an example, for a perfect fluid the flow equation satisfies:

$$\frac{D\mathbf{V}}{\partial s} = \frac{1}{\mu + p}\,{}_{\perp}\partial p\,.$$

Each of the remaining (transverse) orthonormal vectors satisfies the following differential equation:

$$\frac{D_F \mathbf{E}_\alpha}{\partial s} = \frac{D\mathbf{E}_\alpha}{\partial s} + \hat{\gamma}\left(\mathbf{E}_\alpha, \frac{D\mathbf{V}}{\partial s}\right)\mathbf{V} = 0\,. \tag{F.6}$$

These define constant vector fields under the Fermi derivative. These fields can be shown to maintain constant angles with each other, including themselves. Therefore they are in fact orthonormal [orthogonal and unit] vector fields. The set provides a non-rotating frame of reference defined at each point along the streamline $\gamma(s)$. In this basis, along the streamline, the metric is diagonal, with the time component +1 and the space components -1. The derivative of the components of the metric along the streamline vanishes: $\left(d\hat{\gamma}_{\alpha\beta}/ds\right)_{\gamma(s)} = 0$. The gradients transverse to the flow will not in general vanish and have values set by the curvature.

F.3 Deviation using Fermi Derivatives

The first and second order deviation equations can be written in terms of the Fermi derivatives, again without any specific reference to the dimensionality of space–time:

$$\frac{D_{F\perp} Z^\mu}{\partial s} = V^\mu{}_{;\nu\perp} Z^\nu$$

$$\frac{D^2{}_{F\perp} Z^\mu}{\partial s^2} = \left(R^\mu{}_{\nu\sigma\rho} V^\nu V^\sigma + \dot{V}^\mu_{;\rho} - \dot{V}^\mu \dot{V}_\rho\right)_\perp Z^\rho$$

I introduce coefficients Z^α using orthonormal coordinates for the separation $\mathbf{Z} = Z^\alpha \mathbf{E}_\alpha$. Since the Fermi derivative of the coordinate vectors $\{\mathbf{E}_\alpha\}$ vanishes, the Fermi derivative of $\mathbf{Z}$ is given by the ordinary derivative of Z^α with respect to the path length s:

$$
\begin{aligned}
{}_\perp\mathbf{Z} &= {}_\perp Z^a \partial_a = Z^\alpha \mathbf{E}_\alpha \\
\mathbf{V} &= V^\mu \partial_\mu = E_s \\
\frac{dZ^\alpha}{ds} &= V^\alpha{}_{;\beta} Z^\beta \\
\frac{d^2 Z^\alpha}{ds^2} &= \left(R^\alpha{}_{ss\gamma} + \dot{V}^\alpha_{;\gamma} - \dot{V}^\alpha \dot{V}_\gamma \right) Z^\gamma
\end{aligned}
$$

In this basis, the flow is along the streamline $\mathbf{V} = \mathbf{E}_s$. These equations involve ordinary derivatives on the scalar coefficients.

F.4 Evolution Matrix

The first of these is a linear equation, so has a solution of the form:

$$Z^\alpha(s) = A^\alpha{}_\beta(s) Z^\beta\Big|_q .$$

The sum is over the transverse coordinates only. The matrix $A^\alpha{}_\beta(s)$ is the unit matrix at the point q and itself satisfies a linear equation:

$$\frac{dA^\alpha{}_\beta(s)}{ds} = V^\alpha{}_{;\gamma} A^\gamma{}_\beta(s) .$$

Just as I argued that the coefficients Z^α are scalars, I argue that in general the tensor $V^\mu{}_{;\nu} = V^\alpha{}_{;\beta} E^\mu_\alpha E^\beta_\nu$ can be written as the sum of scalars multiplying a tensor product of the orthonormal vectors. Therefore, the second equation, the deviation equation, can be expressed in terms of the matrix as well:

$$\frac{d^2 A^\alpha{}_\beta(s)}{ds^2} = \left(R^\alpha{}_{ss\gamma} + \dot{V}^\alpha_{;\gamma} - \dot{V}^\alpha \dot{V}_\gamma \right) A^\gamma{}_\beta .$$

In any number of dimensions, the matrix $A^\alpha{}_\beta(s)$ can be decomposed into a factor of an orthogonal matrix and a factor of a symmetric matrix:

$$A^\alpha{}_\beta(s) = O^\alpha{}_\gamma(s) S^\gamma{}_\beta(s) .$$

Both factors are unit matrices at some initial point q. The next order terms are anti-symmetric for the orthogonal matrix and symmetric for the symmetric matrix. These considerations give further insight into Eq. (F.3). As mentioned earlier, the anti-symmetric tensor $\omega_{\mu\nu}$ is the ***vorticity tensor*** and the symmetric tensor $\theta_{\mu\nu}$ is the ***expansion tensor***. The ***volume expansion*** is the trace of the expansion tensor in the orthogonal subspace:

$$\theta = \theta_{\mu\nu} h^{\mu\nu} = V^{\mu}{}_{;\mu}.$$

The trace free part of the expansion tensor is the ***shear tensor***:

$$\sigma_{\mu\nu} = \theta_{\mu\nu} - \frac{1}{D-1} h_{\mu\nu}\theta .$$

These identifications provide insight into the equations that result.

The transverse gradient of the flow, as well as the associated expansion and vorticity tensors, can be expressed in terms of the matrix $A^{\alpha}{}_{\beta}$:

$$V^{\alpha}{}_{;\beta} = \frac{dA^{\alpha}{}_{\gamma}}{ds} A^{-1\gamma}{}_{\beta}$$

$$\omega^{\beta}{}_{\alpha} = A^{-1\gamma}{}_{[\alpha} \frac{dA^{\beta]}{}_{\gamma}}{ds}$$

$$\theta^{\beta}{}_{\alpha} = A^{-1\gamma}{}_{(\alpha} \frac{dA^{\beta)}{}_{\gamma}}{ds}$$

$$\theta = \frac{d}{ds} \ln \det A$$

The last equation identifies the volume expansion as related to the differential change of the determinant of the symmetric transformation:

$$\theta = \frac{d}{ds} \ln \det A = \frac{d}{ds} \ln \det S .$$

The deviation equation can now be used to relate the evolution of the vorticity and expansion tensors:

$$\frac{d^2 A^{\alpha}{}_{\gamma}(s)}{ds^2} A^{-1\gamma}{}_{\beta} = R^{\alpha}{}_{ss\beta} + \dot{V}^{\alpha}{}_{;\beta} - \dot{V}^{\alpha}\dot{V}_{\beta} .$$

Except for slight changes in sign due to my metric convention, this is the usual result and so provides the standard set of equations for the evolution of the covariant derivative:

$$\frac{dV_{\alpha;\beta}}{ds} = R_{\alpha ss\beta} + \dot{V}_{\alpha;\beta} - \dot{V}_{\alpha}\dot{V}_{\beta} + V_{\alpha;\gamma}V_{\gamma;\beta}\,. \tag{F.7}$$

I emphasize however that though the form of the result is standard, the space on which this result applies is quite different from the ordinary one of physics. In the numerical examples provided in Sec. 6, I consider seven-dimensional space–time. The above gradients are in a six dimensional sub-space. There are significantly more complex behaviors possible in such a space than the normal three-dimensional space of physics. I identified generalized "cork-screws" as one such set of possibilities in the numerical examples

F.5 Expansion Compression and Vorticity

I consider Eq. (F.7) and separate out the expansion and vorticity components:

$$\begin{aligned} \frac{d\omega_{\alpha\beta}}{ds} &= \dot{V}_{[\alpha;\beta]} - 2\omega_{\gamma[\alpha}\theta_{\beta]\gamma} \\ \frac{d\theta_{\alpha\beta}}{ds} &= \omega_{\alpha\gamma}\omega_{\gamma\beta} + \theta_{\alpha\gamma}\theta_{\gamma\beta} + R_{\alpha ss\beta} + \dot{V}_{(\alpha;\beta)} - \dot{V}_{\alpha}\dot{V}_{\beta} \end{aligned} \,. \tag{F.8}$$

I obtain a coupled set of equations for the vorticity and expansion tensors. In particular I note that once the equations have been solved, I obtain the flow gradients:

$$V_{\mu;\nu} = \theta_{\alpha\beta}E^{\alpha}_{\mu}E^{\beta}_{\nu} + \omega_{\alpha\beta}E^{\alpha}_{\mu}E^{\beta}_{\nu} + \dot{V}_{\alpha}E^{\alpha}_{\mu}E^{s}_{\nu}\,.$$

I can go further in specifying a complete set of linear equations for the special case of a perfect fluid in which the energy density is a function of the pressure (a ***homogeneous*** fluid):

$$\dot{V}_{\alpha} = h^{\beta}_{\alpha}\frac{\partial_{\beta}p}{\mu + p}, \quad \dot{V}_{[\alpha;\beta]} = -\omega_{\alpha\beta}\frac{1}{\mu + p}\frac{dp}{ds}\,.$$

The energy–momentum conservation laws and streamline equations in the co-moving frame are:

$$\frac{dV_\alpha}{ds}=\frac{\partial_\alpha p}{\mu+p},\quad \frac{d\mu}{ds}=-(\mu+p)\theta,\quad \frac{dx^\alpha}{ds}=V^\alpha .$$

These linear differential equations coupled with the linear differential equations of Eq. (F.8) are sufficient to solve the streamline equations. On an initial surface, I assume the pressure, (and hence the energy density) is known. If I evolve the entire surface a small step, the equations indeed determine the next point, since the initial pressure gradients $\partial_\alpha p$ and $p_{;\alpha\beta}$ are also known. For numerical calculations it would be nice to have linear differential equations for these pressure gradients.

I believe the answer is to consider $\partial_\alpha p$ a special case of the transverse vector **Z** discussed above. In this case I expect that the equation for the first order gradient is:

$$\frac{dp_{;\alpha}}{ds}=\left(\theta_\alpha{}^\beta+\omega_\alpha{}^\beta\right)p_{;\beta}.$$

Based on the properties of the Fermi derivative, I expect the second order gradient equation to be:

$$\frac{dp_{;\alpha\beta}}{ds}=\left(\theta_\alpha{}^\gamma+\omega_\alpha{}^\gamma\right)p_{;\gamma\beta}+\left(\theta_\beta{}^\gamma+\omega_\beta{}^\gamma\right)p_{;\alpha\gamma}.$$

I have a complete set of linear partial differential equations allowing the determination of all the variables along a streamline.

I finish by quoting a known result now applied to the dynamic theory of games; for the perfect fluid, the vorticity equation provides the following conservation law along the streamline:

$$\frac{d}{ds}\left(\mathbf{A}^{\mathrm{T}}\boldsymbol{\omega}\mathbf{A}\right)=\frac{1}{\mu+p}\frac{dp}{ds}\left(\mathbf{A}^{\mathrm{T}}\boldsymbol{\omega}\mathbf{A}\right)\equiv-\frac{1}{\lambda}\frac{d\lambda}{ds}\left(\mathbf{A}^{\mathrm{T}}\boldsymbol{\omega}\mathbf{A}\right)$$

$$\lambda\mathbf{A}^{\mathrm{T}}\boldsymbol{\omega}\mathbf{A}=const.$$

$$\mathbf{X}=\mathbf{A}\mathbf{X}^0$$

$$\mathbf{Y}=\mathbf{A}\mathbf{Y}^0$$

$$\lambda X^\alpha\omega_{\alpha\beta}Y^\beta=const.$$

With the modification of the factor λ, the projection of the vorticity tensor onto two vectors stays constant along the streamline and vectors evolve according to the evolution equations. In three dimensions, this says that the projection of the vorticity "vector" $\Omega_a = \varepsilon_{abc}\omega_{bc}$ onto the area $\mathbf{X}\times\mathbf{Y}$ stays constant: The vorticity moves with the fluid[49]. For fluids with viscosity this ceases to be true.

This conservation of vorticity is an appropriate property of a perfect fluid and one that extends to higher dimensions. It is a property I can expect for games as long as I can ignore viscosity. With viscosity I expect a modification to the theory. For three dimensions, the vorticity equation with viscosity has the approximate form:

$$\frac{d\omega_{ab}}{ds} = -\theta_{ac}\omega_{cb} - \omega_{ac}\theta_{cb} + \frac{\eta}{\rho}\nabla^2\omega_{ab}.$$

It is assumed that the fluid is incompressible and that the energy density is approximately equal to the market density. Ignoring the first two terms, the equation is a diffusion equation for the vorticity. The implication for game theory is that a collection of games that play near to an equilibrium strategy should exhibit vorticity. The above equation suggests that the plays will become increasingly fuzzy as time goes on as the vorticity diffuses through choice–space. It will be like a smoke ring that becomes fuzzy as it moves along.

---

[49] See for example Feynman (1963), Volume 2.

Appendix G

Player Fluid

I recall that a perfect fluid provides a specific form for the energy–momentum tensor Eq. (1.4). I note that the energy–momentum tensor can be decomposed into two parts:

- The ***energy density*** $\mu = V^{\mu} T_{\mu\nu} V^{\nu}$ and
- the ***stress tensor*** $p_{\mu\nu} = -h_{\mu\rho} T^{\rho\sigma} h_{\sigma\nu}$.

The projection tensor $h^{\mu\nu} = \hat{\gamma}^{\mu\nu} - V^{\mu} V^{\nu}$ is transverse to the fluid flow. The projection tensor satisfies the condition $\mathbf{hh} = \mathbf{h}$. A perfect fluid has the property that the stress tensor is proportional to this projection tensor: $\mathbf{p} = p\mathbf{h}$.

I propose to generalize the perfect fluid so that it corresponds more intuitively to a number of players engaged in a game. For any geometry there will be an energy density and a flow, which are the unique time-like eigenvalue and eigenvector of the energy–momentum tensor:

$$V_{\mu} T^{\mu\nu} = \mu V^{\nu} . \qquad \text{(G.1)}$$

The generalization of the perfect fluid can be expressed in terms of the stress tensor. I take the defining idea of a ***player*** as the existence of symmetric ***player projection tensors*** $\mathbf{Q}_m$: $\mathbf{Q}_m \mathbf{Q}_n = \delta_{mn} \mathbf{Q}_m$. Not only are these mutually orthogonal but they are orthogonal to the flow. There is one such projection for each player. In the co-moving frame, there is a basis in which each projection tensor $Q^{\mu}{}_{\nu}$ is a "block" identity matrix. The dimensionality of this block is the ***player dimension*** N_m equal to the number of strategies for that player plus one dimension for the value–choice strategy. By this construction it is evident that the projection tensors exist and are mutually orthogonal. I transform these player projection tensors to an arbitrary frame.

I define a ***player fluid*** as one in which the stress tensor is a linear combination of these tensors:

$$\mathbf{p} = \sum_{m \in players} p_m \mathbf{Q}_m \,. \tag{G.2}$$

I term this the ***Player Independence Hypothesis***. It says that in the right basis of the co-moving frame, each player can apply stress only to their choices. It makes the further hypothesis that in the central frame, the pressure and projection tensors depend only on the active coordinates. There will be one independent scalar pressure associated with each player.

To get insight into the form of the projection tensors in the central frame, I start with the equation for the projection tensor:

$$Q^{\mu}{}_{\nu} Q^{\nu}{}_{\lambda} = Q^{\mu}{}_{\lambda} \,.$$

I write the equation for the projection tensor in terms of its active and inactive components in the central frame. For this argument I explicitly assume that time is inactive (there may of course still be other inactive strategies but they are not explicitly displayed):

$$Q^{a}{}_{0} = -\frac{\gamma_{00}}{V_0} Q^{ac} g_{cd} V^d$$

$$Q^{ab} = Q^{a}{}_{c}\left(g^{cd} + W^c W^d\right) Q^{b}{}_{d}$$

$$W^a \equiv \frac{V^a}{\sqrt{1 - g_{cd} V^c V^d}}$$

$$g_{ab} W^a W^b = \frac{g_{ab} V^a V^b}{1 - g_{cd} V^c V^d} = \frac{1 - \gamma^{00} V_0 V_0}{\gamma^{00} V_0 V_0} \equiv -\beta^2 \le 0$$

The first equation is the statement that the projection tensor is orthogonal to the flow. The second statement is the statement that the space components satisfy the projection equation and I have used the first relationship to eliminate the mixed time component. I have simplified the expression by introducing a new flow vector. The factor in the middle of the first equation is the inverse of the projection operator in active space:

$$h^{ac}\left(\delta_c^b + W_c W^b\right) = \left(g^{ac} - V^a V^c\right)\left(\delta_c^b + W_c W^b\right) = g^{ab} \,.$$

I check that the projection equation holds for the mixed space–time and time–time components. I start with the mixed space–time components:

$$Q^a{}_0 = Q^a{}_\nu Q^\nu{}_0 = \gamma^{00} Q_{00} Q^a{}_0 + Q^{ac} g_{cb} Q^b{}_0$$

$$-\frac{\gamma_{00}}{V_0} Q^{ac} g_{cd} V^d = \begin{Bmatrix} -\gamma^{00} \dfrac{\gamma_{00}\gamma_{00}}{V_0 V_0} V^d g_{db} Q^{be} g_{ef} V^f \left(\dfrac{\gamma_{00}}{V_0} Q^{ac} g_{cd} V^d \right) \\ -Q^{ac} g_{cb} \dfrac{\gamma_{00}}{V_0} Q^{bd} g_{de} V^e \end{Bmatrix}.$$

$$Q^{ac} g_{cd} V^d = Q^{ae} \left(g_{ed} W^d g_{fb} W^f + g_{eb} \right) Q^{bc} g_{cd} V^d$$

$$Q^{ac} g_{cd} V^d = Q^{ac} g_{cd} V^d$$

This shows that the mixed space–time component is consistent with the above equation for the space–space components. Similarly I check the time–time component:

$$Q_{00} = Q_{0\nu} Q^\nu{}_0 = Q_{00} Q^0{}_0 + Q_{0a} Q^a{}_0$$

$$Q_{00} = \gamma^{00} Q_{00} Q_{00} + g_{ab} Q^a{}_0 Q^b{}_0$$

$$V^d g_{dc} Q^{ce} g_{ef} V^f = V^d g_{dc} Q^{ca} \left(g_{ah} W^h W^g g_{gb} + g_{ab} \right) Q^{be} g_{ef} V^f \cdot$$

$$V^d g_{dc} Q^{ce} g_{ef} V^f = V^d g_{dc} Q^{ce} g_{ef} V^f$$

I summarize the result that the constraints on the projection tensor are specified by the active component equation:

$$Q^{ab} = Q^a{}_c \left(g^{cd} + W^c W^d \right) Q^b{}_d \,. \tag{G.3}$$

This has some but not all of the constraints for the given player projection tensor. It does not for example give the player dimension of the tensor. To see that I would need to return to the co-moving frame. More generally, to consider the constraints on the stress tensor, I start in the co-moving frame with the diagonal matrix $p^\mu{}_\nu$. As per the player independence hypothesis, this is a matrix specified by the pressure scalars for each player.

From the stress tensor I compute the eigenvalue equation:

$$char(\lambda) = \det(\lambda \mathbf{I} - \mathbf{p}) = 0 \,.$$

For each player, there will be a set of identical eigenvalues equal to the scalar pressure for that player. The number of such eigenvalues equals the player dimension N_m. From this eigenvalue equation specified by the scalar pressures, I construct the invariant characteristic equation for the stress tensor in any frame:

$$char(\mathbf{p}) = 0 . \tag{G.4}$$

I count the number of variables and independent equations: I first count the variables:

Type	Variables	Number	Comments
Flow	V^μ	$D-1$	Unit Vector
Energy	μ	1	Eigenvalue
Stress Tensor	$p^{\mu\nu}$	$\frac{1}{2}D(D+1)$	Symmetric Matrix
Characteristic Coefficients	$char(\mathbf{p})$	$D-1$	Transverse Only
Total		$\frac{1}{2}D(D+3)+D-1$	

I next count the equations and see that there is an excess of $D-1$ variables over the number of equations, which can be ascribed to the number of unknown coefficients of the characteristic equation:

Equations	Number	Comments
$d\mu/ds + \theta\mu + \theta_{\mu\nu}p^{\mu\nu} = 0$	1	Eq. (G.11)
$\left(\mu h^{\mu\lambda} + p^{\mu\lambda}\right)\dot{V}_\lambda = h^\mu_\rho h^\sigma_\lambda p^{\rho\lambda}{}_{;\sigma}$	$D-1$	Eq. (G.13)
$p^{\mu\nu}V_\nu = 0$	D	Stress is orthogonal to flow
$char(\mathbf{p})^{\mu\nu} = 0$	$\frac{1}{2}D(D-1)$	Symmetric Characteristic equation
Total	$\frac{1}{2}D(D+3)$	

If there were a single unknown coefficient in the characteristic equation, which is specified for example by assuming that the stress is proportional to the flow projection, then the excess of variables over equations is one. I specified that by the homogeneity assumption that the energy density is a function of the pressure.

For two players each with a distinct pressure, the excess is two and for N players the excess is N. Given these coefficients of the characteristic equation, the energy density and flow variables are determined.

In deriving the first two rows of the table of independent equations, I stated a result for the conservation of energy and momentum. I now say a little more about those results. In the central frame I decompose the stress tensor into the active and inactive choices. I consider any general second order symmetric tensor $Q_{\mu\nu}$. I express the components of this tensor in terms of gauge invariant quantities:

$$\left\{ Q^{ab} \quad Q_{jk} \quad Q^{a}{}_{j} \right\}. \tag{G.5}$$

The upper gauge dependent quantities are:

$$\begin{aligned} Q^{ak} &= \gamma^{kj} Q^{a}{}_{j} - A^{k}_{b} Q^{ab} \\ Q^{kj} &= \gamma^{jl} \gamma^{ki} Q_{il} + A^{j}_{b} A^{k}_{c} Q^{bc} - \gamma^{jl} A^{k}_{b} Q^{b}{}_{l} - \gamma^{kl} A^{j}_{b} Q^{b}{}_{l} \end{aligned}. \tag{G.6}$$

The lower gauge dependent quantities are:

$$\begin{aligned} Q_{aj} &= g_{ab} Q^{b}{}_{j} + A^{k}_{a} Q_{kj} \\ Q_{ca} &= Q^{db} g_{cd} g_{ba} + A^{j}_{c} A^{k}_{a} Q_{jk} + g_{cb} A^{j}_{a} Q^{b}{}_{j} + A^{j}_{c} g_{ab} Q^{b}{}_{j} \end{aligned}. \tag{G.7}$$

And the mixed gauge dependent quantities are:

$$\begin{aligned} Q^{j}{}_{k} &= \gamma^{jl} Q_{lk} - A^{j}_{a} Q^{a}{}_{k} \\ Q^{a}{}_{b} &= Q^{ac} g_{cb} + A^{k}_{b} Q^{a}{}_{k} \\ Q^{k}{}_{a} &= g_{ab} \gamma^{kj} Q^{b}{}_{j} - A^{j}_{a} A^{k}_{b} Q^{b}{}_{j} - g_{ab} A^{k}_{c} Q^{bc} + \gamma^{kl} A^{j}_{a} Q_{lj} \end{aligned}. \tag{G.8}$$

Thus all the components of the tensor are determined by the gauge independent components. In particular, the above relationships hold for the stress tensor $p^{\mu\nu}$. In addition, the stress tensor is orthogonal to the flow:

$$\begin{aligned} \gamma^{jk} V_{j} p^{a}{}_{k} + p^{ab} g_{bc} V^{c} &= 0 \\ \gamma^{jl} V_{j} p_{lk} + g_{ac} V^{c} p^{a}{}_{k} &= 0 \end{aligned}. \tag{G.9}$$

These orthogonality relationships are expressed in terms of the gauge independent components. I have used the special case of this earlier where time was the only explicit inactive variable:

$$\begin{aligned} p^{a}{}_{0} &= -\frac{\gamma_{00}}{V_0} p^{ab} g_{bc} V^c \\ p_{00} &= \frac{\gamma_{00}\gamma_{00}}{V_0 V_0} V^a g_{ab} p^{bc} g_{cd} V^d \end{aligned} \qquad \text{(G.10)}$$

With these relationships, I write the conservation laws $T^{\mu\nu}{}_{;\nu} = 0$ in terms of the active and inactive coordinates, starting with the longitudinal components:

$$\frac{d\mu}{ds} + \theta\mu + \theta_{\mu\nu} p^{\mu\nu} = 0 . \qquad \text{(G.11)}$$

The product of the stress and compression can be expanded in terms of the active and inactive components:

$$\theta_{\mu\nu} p^{\mu\nu} = p_{il}\gamma^{ik}\gamma^{lj}\theta_{kj} + p^{ab} g_{ac} g_{bd} \theta^{cd} + 2\gamma^{jk} g_{ab} p^{a}{}_{j} \theta^{b}{}_{k} . \qquad \text{(G.12)}$$

The transverse components can be shown to be given by

$$\left(\mu h^{\mu\lambda} + p^{\mu\lambda}\right)\dot{V}_{\lambda} = h^{\mu}_{\rho} h^{\sigma}_{\lambda} p^{\rho\lambda}{}_{;\sigma} . \qquad \text{(G.13)}$$

It provides an intuitive generalization to the perfect fluid. It can be expressed in terms of the active and inactive components. I write first the active components:

$$\mu V^{a}{}_{|b} V^{b} = \begin{pmatrix} g^{ab} F^{k}_{bc} \left(\mu V^c V_k - p^{c}{}_{k}\right) \\ -\tfrac{1}{2}\left(\mu g^{ab} V_j V_k + p^{ab}\gamma_{jk} - g^{ab} p_{jk}\right)\partial_b \gamma^{jk} \\ + p^{ab}{}_{|b} \\ + V^a \left(p_{il}\gamma^{ik}\gamma^{lj}\theta_{kj} + p^{bc} g_{bd} g_{ce} \theta^{de} + 2\gamma^{jk} g_{bc} p^{b}{}_{j} \theta^{c}{}_{k}\right) \end{pmatrix} . \qquad \text{(G.14)}$$

Next I write the conservation laws for the inactive components:

$$\begin{aligned} &\mu \frac{dV_i}{ds} - \left(p_{jl}\gamma^{jk}\gamma^{lm}\theta_{km} + p^{ab} g_{ac} g_{bd} \theta^{cd} + 2\gamma^{jk} g_{ab} p^{a}{}_{j} \theta^{b}{}_{k}\right) V_i \\ &= \frac{1}{\sqrt{g\gamma}} \partial_b \left(\sqrt{g\gamma}\, p^{b}{}_{i}\right) \end{aligned} . \qquad \text{(G.15)}$$

These equations are the natural generalization to the perfect fluid. They indicate how the various components of the stress tensor influence the rate of change of flow. I thus see that the characteristic equation, the equations that reflect that the stress is orthogonal to flow and the conservation laws determine stress tensor up to N scalar functions.

The metric is assumed known; it will in fact be determined by the full set of field equations. These equations will have changed because of the new assumed form for the stress tensor. I look at one such example and consider the impact of the player fluid on the source Eq. (C.1):

$$\tfrac{1}{2}\frac{1}{\sqrt{|g\gamma|}}\partial_b\left(\sqrt{|g\gamma|}g^{ac}g^{bd}\gamma_{jk}F^k_{cd}\right)=\kappa\left(\mu V^a V_j - p^a{}_j\right). \qquad \text{(G.16)}$$

As before, the right hand side is a conserved current, which follows from the form on the left. The impact of a player fluid will be the existence of a dependency on the mixed stress component. The player independence hypothesis makes a significant modification on the sources for the market field.

Bibliography

Birkhoff, G., and MacLane, S., (1959), *A Survey of Modern Algebra, Revised Edition*, (The Macmillan Company, New York) 45

Brooks, F.P., (1982), *The Mythical Man Month*, (Addison–Wesley, Massachusetts) 143

Chandrasekhar, S., (1961), *Hydrodynamic and Hydromagnetic Stability*, (Dover Publications, New York) 163

Chevalley, C., (1946), *Theory of Lie Groups*, (Princeton University Press, Princeton) 167

Club of Rome: Meadows, D.H., Meadows, D.L., Randers, J., and Behrens, W.W., (1972), *The Limits to Growth*, (Signet Book, New York) vii

Crawford, W.P., (1992), *Mariner's Weather*, (W.W. Norton, New York) ..passim

Davies, P.C.W. and Brown, J., (1988), *Superstrings: A Theory of Everything?*, (Cambridge University Press, Cambridge). 15

DeMarco, T., (1982), *Controlling software projects*, (Yourdon Press, New York) 143

Devaney, Robert L., (1989), *An Introduction to Chaotic Dynamical Systems, Second Edition*, (Addison-Wesley Publishing Company, Inc., Redwood City) 41

Dresher, M., (1981), *The Mathematics of Games of Strategy*, (Dover Publications, New York) 2

Eatwell, J., Milgate, M., and Newman, P., (1987), *Game Theory*, (W.W. Norton, New York) 2

Einstein, A., Lorentz, H.A., Minkowski, H., Weyl, H., (1952), *The Principle of Relativity: A Collection of original memoirs on the special and general theory of relativity, with notes by Sommerfeld, A.*, Translated by Perrett, W., and Jeffrey, G.B. (Dover Publications, Toronto) 6

Friedman, R.M., (1989), *Appropriating the Weather, Vilhelm Bjerknes and the construction of modern meteorology*, (Cornell University Press, Ithaca) 44
Göckeler, M., and Schüker, T., (1987), *Differential Geometry, Gauge Theories, and Gravity*, (Cambridge Universiy Press, Cambridge) 33, 54, 57, 58
Green, M. B., Schwarz, J.H. and Witten, E., (1987), *Superstring theory*, (Cambridge University Press, Cambridge). 15
Hamermesh, M., (1962), *Group Theory and its Application to Physical Problems*, (Addison–Wesley, Reading, MA).. 167
Hawking, S. W., and Ellis, G.F.R., (1973), *The large scale strutructure of space-time*, (Cambridge University Press, Cambridge).................................. passim
High Performance Systems (1997), *ithink® Software* 51, 96
Landau, L.D., Lifshitz, E.M., (1958), *Statistical Physics*, (Addison–Wesley, Reading, MA) 199
Letcher, J.S. Jr., (1977), *Self-Contained Celestial Navigation with H.O. 208*, (International Marine Publishing Company, Camden, Maine) 68
Luce, R.D., and Raiffa, H. (1957), *Games and Decisions*, (Dover Publications, New York) .... passim
Mas-Colell, A., Whinston, M. D., and Green, J. R., (1995), *Macroeconomic Theory*, (Oxford University Press, Oxford)... 2
Myerson, R. B. , (1991), *Game Theory: Analysis of Conflict*, (Harvard University Press, Cambridge, MA).. 2
Ordeshook, P.C., (1986), *Game Theory and Political Theory: An Introduction*, (Cambridge University Press, Cambridge) 2
Osborne, M.J., and Rubinstein, A., (1994), *A Course in Game Theory*, (The MIT Press, Cambridge, MA)... 2
Oxford English Dictionary, (2002), (Oxford University Press, Oxford) ... 39, 157
Pauli, W., (1958), *Theory of Relativity*, (Pergamon Press, New York)........ 15, 60
Pippard, A.B., (1961), *Classical Thermodynamics*, (Cambridge University Press, London).. 194, 195, 199
Rorty, R., (1989), *Contingency, Irony, and Solidarity*, (Cambridge University Press, Cambridge) 2, 32
Rorty, R., (1991), *Objectivity, Relativism, and Truth*, (Cambridge University Press, Cambridge) 2
Senge, P.M., (1990), *The Fifth Discipline: the art and practice of the learning organization*, (Currency Doubleday, New York).. vii
Shubik, M., (1991), *Game Theory in the Social Sciences: Concepts and Solutions*, (The MIT Press, Cambridge, MA) 2
Sommerfeld, A., *Partial Differential Equations in Physics*, (1964), [Academic Press, New York] 216
Sun Tzu, (1988), *The Art of War*, Translated by Thomas Cleary, (Shambala, Boston)... 1, 99
Synge, J.L., and Schild, A., (1949), *Tensor Calculus*, (Dover Publications, New York).. 33, 43

Tolman, R.C., (1987), *Relativity, Thermodynamics, and Cosmology*, (Dover Publications, New York).. 157, 161
Von Neumann, J. and Morgenstern, O., (1944), *Theory of Games and Economic Behavior*, (John Wiley and Sons, New York).. passim
Weyl, H., (1922), *Space–Time–Matter*, Translated from the German by Brose, H.L., (Dover Publication, USA).. 15
Williams, J.D., (1966), *The Compleat Strategyst*, (McGraw-Hill, New York).... 2
Winograd, T., and Flores, F., (1986), *Understanding Computers and Cognition*, (Addison–Wesley Publishing, Reading).. 2, 17, 60
Wolfram, S., (1992), *Mathematica: A System for Doing the Mathematics by Computer*, Second Edition, (Addison–Wesley, Reading, MA)............. 96, 209
Wolstenholme, E.F., (1990), *System Enquiry*, (John Wiley and Sons, New York).. vii

Index